Lecture Notes in Geoinformation and Cartography

For further volumes:
http://www.springer.com/series/7418

Alias Abdul Rahman · Pawel Boguslawski
Christopher Gold · Mohamad Nor Said
Editors

Developments in Multidimensional Spatial Data Models

Springer

Editors
Alias Abdul Rahman
Pawel Boguslawski
Christopher Gold
Mohamad Nor Said
Faculty of Geoinformation and Real Estate
Universiti Teknologi Malaysia
Skudai
Malaysia

ISSN 1863-2246 ISSN 1863-2351 (electronic)
ISBN 978-3-662-52389-6 ISBN 978-3-642-36379-5 (eBook)
DOI 10.1007/978-3-642-36379-5
Springer Heidelberg New York Dordrecht London

Softcover reprint of the hardcover 1st edition 2013

Printed on acid-free paper

Springer is part of Springer Science+Business Media (www.springer.com)

Preface

First of all, we would like to thank all the authors who submitted their full papers to the GeoAdvances 2012 workshop, which was held from 7 to 8 November at the Faculty of Geoinformation and Real Estate, Universiti Teknologi Malaysia, Johor Bahru, Johor, Malaysia. All the submitted full papers (42 of them) were blind peer-reviewed by the Scientific Program Committee members, and the 16 accepted papers are published in this Springer book of the Lecture Notes on Geoinformation and Cartography (LNG&C) series. The book discusses recent developments, in multidimensional spatial data modeling, and contains 16 chapters. Each chapter has been thoroughly reviewed by at least three people from the committee and the scored average marks determined the acceptability of the paper. We are very pleased with the quality of the result, and hope that these recent developments in multidimensional spatial data modeling add significant geospatial science knowledge to the GIS community, especially for researchers and professionals in the industry.

The successful completion of this book required the cooperation and understanding of several key individuals. We thank the GeoAdvances 2012 secretariat especially Ms. Latifah Ibrahim for her endless help, the 3D GIS Research Lab postgraduate students, and our understanding wives Faridah, Patrycja, Valerie, and Hasnah.

Alias Abdul Rahman
Pawel Boguslawski
Christopher Gold
Mohamad Nor Said
Chairs, GeoAdvances 2012

Organization

Programme Committee

Alias Abdul Rahman
Pawel Boguslawski
Christopher Gold
Mohamad Nor Said

Reviewers

Albert Chong—The University of Southern Queensland, Australia
Clive Fraser—University of Melbourne, Australia
Fabio Remondino—Bruno Kessler Foundation, Italy
François Anton—Technical University of Denmark, Denmark
Gerhard Gröger—University of Bonn, Germany
Giles Foody—University of Nottingham, United Kingdom
Hugo Ledoux—Delft University of Technology, The Netherlands
Ismail Rakip Karas—Karabuk University, Turkey
Jane Drummond—University of Glasgow, United Kingdom
Jantien Stoter—Delft University of Technology, The Netherlands
Jay Hyoun Kwon—University of Seoul, Korea
Jiyeong Lee—University of Seoul, Korea
Kazimierz Becek—Universiti Brunei Darussalam, Brunei
Lars Bodum—Aalborg University, Denmark
Md Latifur Rahman Sarker—Universiti Teknologi Malaysia, Malaysia
P. C. Lai—The University of Hong Kong, Hong Kong
Peter van Oosterom—Delft University of Technology, The Netherlands
Rahmi Nurhan Celik—Istanbul Technical University, Turkey
Shuanggen Jin—Chinese Academy of Sciences, China
Sisi Zlatanova—Delft University of Technology, The Netherlands
Tajul Ariffin Musa—Universiti Teknologi Malaysia, Malaysia

Umit Isikdag—Beykent University, Turkey
Volker Coors—Stuttgart University of Applied Science, Germany
Xiaoli Ding—The Hong Kong Polytechnic University, Hong Kong

Additional Reviewers

Filip Biljecki—Delft University of Technology, The Netherlands
Junqiao Zhao—Delft University of Technology, The Netherlands
Zulkepli Majid—Universiti Teknologi Malaysia, Malaysia
Mohd Nadzri Md Reba—Universiti Teknologi Malaysia, Malaysia

Local Organizing Committee

Ali Jamali—Universiti Teknologi Malaysia, Malaysia
Behnam Alizadehashrafi—Universiti Teknologi Malaysia, Malaysia
Chengxi Bernad Siew—Universiti Teknologi Malaysia, Malaysia
Dewi Narty Mohd. Ikhsan—Universiti Teknologi Malaysia, Malaysia
Edward Duncan—Universiti Teknologi Malaysia, Malaysia
Khairul Hafiz Sharkawi—Universiti Teknologi Malaysia, Malaysia
Latifah Ibrahim—Universiti Teknologi Malaysia, Malaysia
Nur Amalina Zulkifli—Universiti Teknologi Malaysia, Malaysia
Siddique Ullah Baig—Universiti Teknologi Malaysia, Malaysia
Sarybanon Abd. Ghani—Universiti Teknologi Malaysia, Malaysia

Contents

Medial Axis Approximation of River Networks for Catchment Area Delineation

Farid Karimipour, Mehran Ghandehari and Hugo Ledoux

Abstract The hydrological catchment areas are commonly extracted from digital elevation models (DEMs). The shortcoming is that computations for large areas are very time consuming and even may be impractical. Furthermore, the DEM may be inaccessible or in a poor quality. This chapter presents an algorithm to approximate the medial axis of river networks, which leads to catchment area delineation. We propose a modification to a Voronoi-based algorithm for medial axis extraction through labeling the sample points in order to automatically avoid appearing extraneous branches in the media axis. The proposed approach is used in a case study and the results are compared with a DEM-based method. The results illustrate that our method is stable, easy to implement and robust, even in the presence of significant noises and perturbations, and guarantees one polygon per catchment.

Keywords Voronoi diagram · Delaunay triangulation · Medial axis · River network · Catchment area delineation

1 Introduction

A catchment area is a hydrological unit where precipitations that fall into this area, eventually end up in the same river. The raster-based algorithms, which usually use DEM in their approach, are very common in automated catchment area

F. Karimipour (✉) · M. Ghandehari
Department of Surveying and Geomatics Engineering, College of Engineering, University of Tehran, Tehran, Iran
e-mail: fkarimipr@ut.ac.ir

M. Ghandehari
e-mail: ghandehary@ut.ac.ir

H. Ledoux
OTB, Section GIS Technology, Delft University of Technology, Delft, The Netherlands
e-mail: h.ledoux@tudelft.nl

A. Abdul Rahman et al. (eds.), *Developments in Multidimensional Spatial Data Models*, Lecture Notes in Geoinformation and Cartography,
DOI: 10.1007/978-3-642-36379-5_1,

delineation (Martz and Garbrecht 1992, 1993; Turcotte et al. 2001; Chorowicz et al. 1992; Mark 1984; Tarboton 1997; Lin et al. 2006; Yang et al. 2010; Nelson et al. 1994; Mower 1994; Jones et al. 1990). Nevertheless, they have some difficulties: Firstly, the raster-based analyses can be time consuming. Secondly, the accuracy of these methods depends on the quality and type of the DEM used, that is, *per se*, affected by the accuracy, density and distribution of the source data, the smoothness of the terrain surface and the deployed interpolation method (Li et al. 2005). Thirdly, the limitations in flow direction computation affect the results. The flow path is biased to the grid axes and the water is trapped in sinks and flat areas. Finally, the raster to vector conversion of boundaries may create some intersections and the boundaries are not explicitly defined.

On top of these, the river network may be the only available data in some cases, and the DEM is either inaccessible or in a poor quality. In this situation, the best approximation of the catchment areas would be to extract the region that is closer to a certain river segment than to any other and consider it as the catchment area corresponding to that river (Gold and Dakowicz 2005; Dillabaugh 2002; McAllister 1999). This approximation is achieved through medial axis (MA) extraction, which is defined as the set of points that are equidistant from at least two points on the boundary of the shape.

This chapter aims to verify the hypothesis stated by Gold and Snoeyink (Gold and Snoeyink 2001) to use a Voronoi-based MA extraction method (called one-step crust and skeleton algorithm) for delineation of catchment areas: The river network is sampled with a set of points, and Delaunay triangulation and Voronoi diagrams are used to extract the MA, which results in an approximation of the catchment areas. The initial investigation shows that this method gives a fair approximation of the catchments. However, there are many extraneous branches in the extracted MA (the catchment area, here) due to small perturbations of the sample points.

This chapter proposes a modification to the one-step crust and skeleton algorithm to overcome the above issue through labeling the sample points. Each river segment is considered as a curve segment and its corresponding sample points are assigned the same label. Furthermore, we explain how the catchment polygons are constructed from the extracted MA, which is a set of lines. The conceptual structure and the results illustrate that our method is stable, easy to implement and robust, even in the presence of significant noise and perturbations and guarantees one polygon per catchment.

The rest of the chapter is organized as follows: Sect. 2 represents some geometric definitions, including Delaunay triangulation, Voronoi diagram, medial axis and two definitions related to sampling. Section 3 describes the one-step crust and skeleton algorithm. Simplification and pruning methods for filtering the MA are introduced briefly in Sect. 4. In Sect. 5, we propose a method for catchment area delineation through a modification to the algorithm presented in Sect. 3. The proposed method is used in a case study in Sect. 6 and the results are compared with a DEM-based method. Finally, Sect. 7 discusses the concluding remarks.

2 Geometric Definition

This section represents some geometric preliminaries, including Delaunay triangulation, Voronoi diagram and medial axis. Two definitions related to sampling are presented. In this section, $\mathcal{O}$ is a 2D object, $\partial\mathcal{O}$ is its boundary and $S \subset \partial\mathcal{O}$ is a dense sampling of $\partial\mathcal{O}$.

2.1 Delaunay Triangulation

Definition 1 Given a point set S in the plane, the *Delaunay triangulation* (DT) is a unique triangulation (if the points are in general position) of the points in S that satisfies the circum-circle property: the circum-circle of each triangle does not contain any other point $s \in S$ (Ledoux 2006) (Fig. 1a).

2.2 Voronoi Diagram

Definition 2 Let S be a set of points in R^2. The Voronoi cell of a point $p \in S$, denoted as $V_p\ (S)$, is the set of points $x \in R^2$ that are closer to p than to any other point in S:

$$V_p(S) = \{x \in R^2 | \|x - p\| \leq \|x - q\|, q \in S, q \neq p\} \tag{1}$$

The union of the Voronoi cells of all points $s \in S$ forms the *Voronoi diagram* of S, denoted as $VD(S)$:

$$VD(S) = \bigcup V_p(S), p \in S \tag{2}$$

Figure 1b shows the Voronoi diagrams of a set of points in the plane. Delaunay triangulation and Voronoi diagram are dual structures: the centers of circum-circles of the Delaunay triangles are the Voronoi vertices; and joining the adjacent

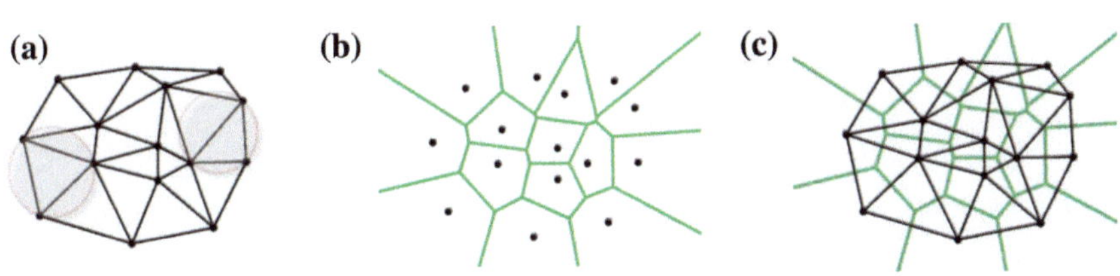

Fig. 1 **a** Delaunay triangulation and **b** Voronoi diagram of a set of points in the plane; and **c** their duality

generator points in the Voronoi diagram yields their Delaunay triangulation (Fig. 1c) (Karimipour et al. 2010).

For Voronoi diagram of sample points S, the Voronoi vertices are classified into *inner* and *outer vertices*, which lie inside and outside $\mathcal{O}$, respectively. Then, the Voronoi edges are classified into three groups: edges between two inner vertices (*inner Voronoi edges*), edges between two outer vertices (*outer Voronoi edges*), and edges between an inner and an outer vertices (*mixed Voronoi edges*).

2.3 Medial Axis

The medial axis was first introduced by Blum (1967) as a tool in image analysis. Grassfire model is the most popular definition of the MA with an intuitive concept: consider starting a fire on the boundary of a shape in the plane. The fire starts at the same moment, everywhere on the boundary and it propagates with homogeneous velocity in every directions. The MA is the set of points where the front of the fire collides with itself, or other fire front. Mathematically:

Definition 3 The *medial axis* is (the closure of) the set of points in $\mathcal{O}$ that have at least two closest points on the object's boundary $\partial\mathcal{O}$. (Amenta et al. 1998).

Another description defines the medial axis as the centers of the set of maximal disks contained in $\mathcal{O}$ (a maximal disk is a disk contained in a shape that is not completely covered by another disk contained in the shape) (Fig. 2).

2.4 Local Feature Size and r-Sampling

Quality of the sample points S has a direct effect on the extracted MA. *Local feature size* is a quantitative measure to determine the level of details at a point on a curve, and the sampling density needed for curve reconstruction and MA extraction.

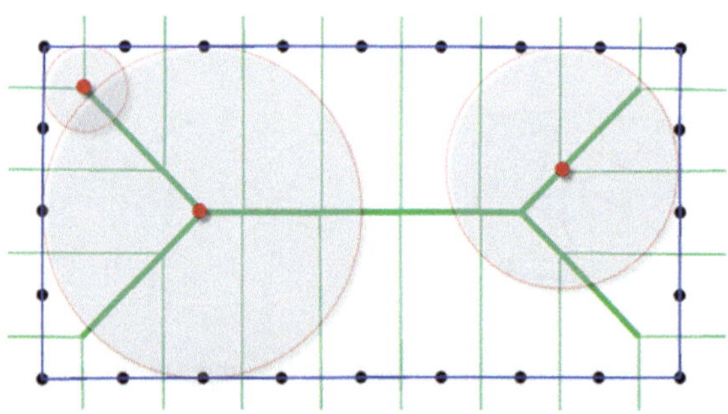

Fig. 2 The MA of a 2D curve (*rectangle*)

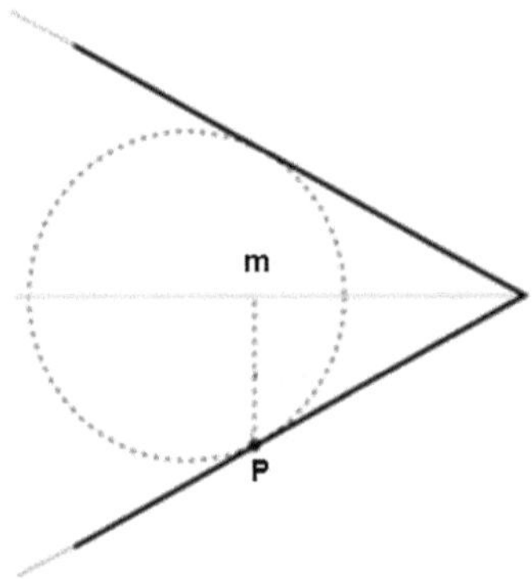

Fig. 3 The local feature size of a point p (*line segment pm*) (Wenger 2003)

Definition 4 The *local feature size* of a point $p \in \partial \mathcal{O}$, denoted as $LFS(p)$, is the distance from p to the nearest point m on the MA (Amenta et al. 1998) (Fig. 3).

Definition 5 The object $\mathcal{O}$ is *r-sampled* by a set of sample points S if for each point $p \in \partial \mathcal{O}$, there is at least one sample point $s \in S$ that $\| p\text{-}s \| \leq r \ast LFS(p)$ (Amenta et al. 1998).

The value of r is less than 1; and usually $r = 0.4$ is considered a reasonably dense sampling (Amenta et al. 1998). Figure 4 shows an example where sample points around the center are denser in order to provide a proper sampling.

3 One-Step Crust and Skeleton Algorithm

Amenta et al. (1998) proposed a Voronoi-based algorithm (called crust algorithm) to reconstruct the boundary from a set of sample points forming the boundary of a shape. Gold and Snoeyink (2001) improved this algorithm so that both the boundary (crust) and the MA (skeleton) are extracted, simultaneously and the coined the name "one-step crust and skeleton" for this algorithm.

In the one-step crust and skeleton algorithm every Voronoi/Delaunay edge is either part of the crust (Delaunay) or the skeleton (Voronoi), which can be determined by a simple *inCircle* test. Each Delaunay edge (D_1D_2 in Fig. 5a) belongs to two triangles ($D_1D_2D_3$ and $D_1D_2D_4$ in Fig. 5a). For each Delaunay edge, there is a dual Voronoi edge (V_1V_2 in Fig. 5a).

Suppose two triangles $D_1D_2D_3$ and $D_1D_2D_4$ have a common edge D_1D_2 whose dual Voronoi edge is V_1V_2. The *InCircle*(D_1, D_2, V_1, V_2) determines the position of V_2 respect to the circle passes through D_1, D_2 and V_1. If V_2 is outside the circle, D_1D_2 belongs to the crust (Fig. 5b). If V_2 is inside, however, V_1V_2 belongs to the skeleton (Fig. 5c).

As illustrated in Fig. 6, if the one-step crust and skeleton algorithm is applied on a river network, the extracted MA yields a fair approximation of the

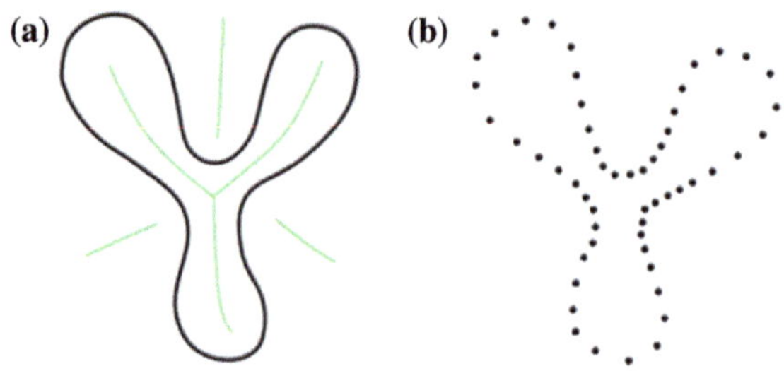

Fig. 4 **a** A curve with its MA (*green curves*); **b** An r-sampling of the curve (Wenger 2003)

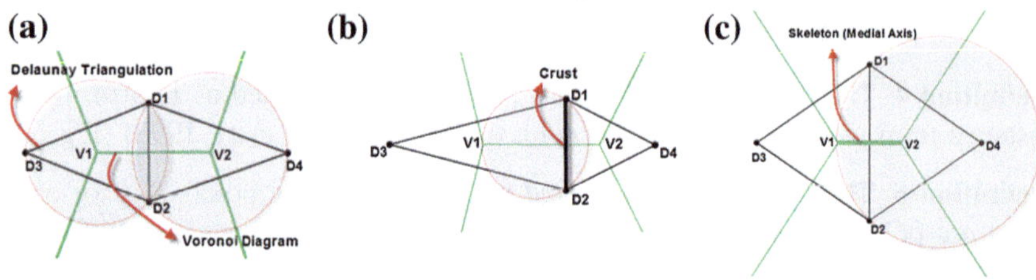

Fig. 5 One-step crust and skeleton extraction algorithm: **a** Delaunay triangulation and Voronoi diagram of four sample points D_1 to D_4; **b** V_2 is outside the circle passes through D_1, D_2 and V_1, so D_1D_2 belongs to the crust; **c** V_2 is inside the circle passes through D_1, D_2 and V_1, so V_1V_2 belongs to the skeleton

Fig. 6 MA extraction of a river network, which yields an approximation of the catchments

corresponding catchments. However, there are many extraneous branches in the extracted MA due to small perturbations of the sample points, which must be filtered out as discussed in the next section.

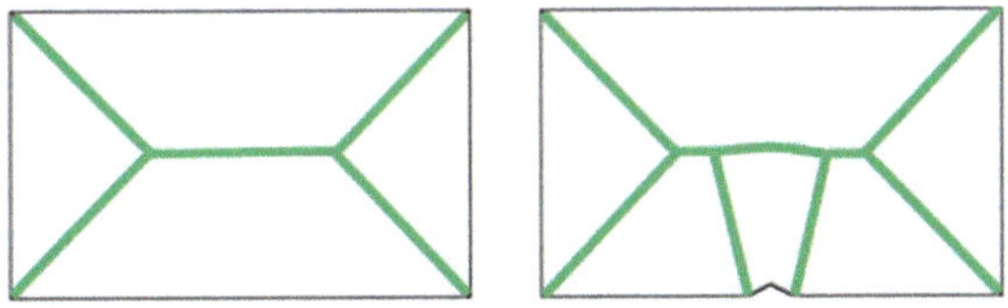

Fig. 7 Similar shapes may have significantly different MAs in the presence of boundary perturbations

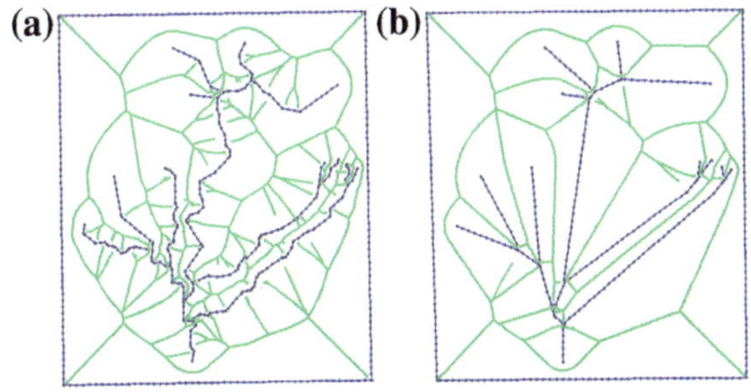

Fig. 8 MA approximation of a river network: **a** before and **b** after simplification

4 Filtering the Extraneous Edges in the Medial Axis

The MA is inherently unstable under small perturbations; i.e., it is very sensitive to the small changes of the boundary, which produce many extraneous branches in the MA. As a result, two similar shapes may have significantly different MAs (Fig. 7). Filtering extraneous branches is a common approach to extract the major parts of the MA. Such filtering may be applied as pre-processing through simplifying (smoothing) the boundary, or as post-processing through pruning that eliminate the extraneous branches of the extracted MA.

4.1 Simplification

Some of the filtering methods simplify the boundary before computing the MA by removing perturbations or boundary noises (Siddiqi et al. 2002; Attali and Montanvert 1996) (Fig. 8). Although these methods aim to remove unwanted boundary noises, they may not provide the ideal results: (1) Some extraneous edges may still exist in the extracted MA; (2) the distinction between boundary data and noise can be difficult; and (3) these methods can alter the topological structure and thus the MA position.

4.2 Pruning

The purpose of the pruning algorithms, as a post-processing step, is to remove extraneous branches of the extracted MA, in order to preserve only the stable parts of the MA. An importance value (based on angle, distance, area, etc.) is assigned to each branch, and then the branches with the importance value less than a given threshold are removed (Attali et al. 1995; Chazal and Lieutier 2005; Attali and Montanvert 1994; Giesen et al. 2009).

The pruning algorithms have some drawbacks: (1) Some extraneous branches may not be eliminated; (2) Eliminating extraneous branches usually shorten the main branches; (3) A disconnection in the main structure of the MA may occur; (4) Most of the pruning methods do not preserve the topology of complex shapes; (5) In some cases, multiple parameters are required and it is difficult to determine appropriate thresholds, simultaneously. Finally, most pruning methods require user checks at the end.

5 Proposed Algorithm for Medial Axis Extraction

In this section, we propose an improvement to the one-step crust and skeleton algorithm through labeling the sample points in order to automatically avoid appearing extraneous branches in the MA, and show how our proposed approach improves the results (Karimipour and Ghandehari 2012; Ghandehari and Karimipour 2012).

Figure 9a illustrates the MA of a shape extracted using the one-step crust and skeleton algorithm. As this figure shows, this algorithm detects some extraneous edges as parts of the MA, which are filtered through simplification or pruning. We observed that such extraneous edges are the Voronoi edges created between the sample points that lie on the same segment of the curve. It led us to the idea of labeling the sample points in order to automatically avoid appearing such edges in the MA (Fig 9b).

We consider the shape boundary as different curve segments ∂_i such that:

$$\partial O = \bigcup_{i=1}^{n} \partial O_i \tag{3}$$

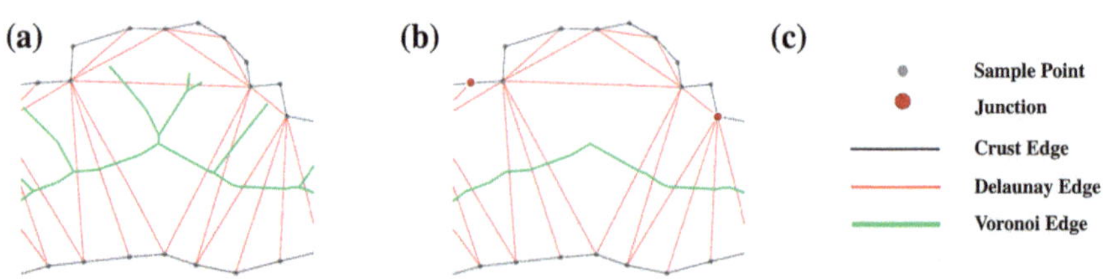

Fig. 9 **a** One-step crust and skeleton algorithm detects some extraneous edges as parts of the MA. They are the Voronoi edges created between the sample points that lie on the same segment of the curve; **b** our proposed method automatically avoid such edges in the MA

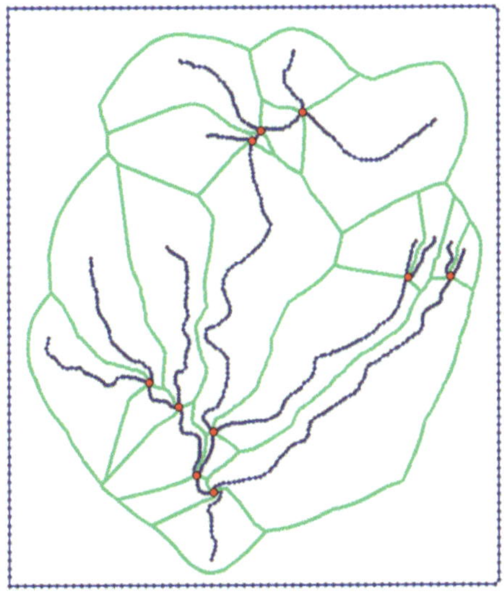

Fig. 10 MA approximation of river network using our proposed algorithm avoids appearing the extraneous branches in the MA

Inner and outer Voronoi edges do not intersect with ∂, but mixed Voronoi edges do (Giesen et al. 2007). The same applied to the Delaunay edges: Delaunay edges of the sample points S are classified into three classes: *Mixed Delaunay edges* that join two consecutive points and belong to the crust; and *inner/outer Delaunay edges* that join two non-consecutive points and are completely inside/outside $\mathcal{O}$ (all Delaunay vertices lie on the $\partial\mathcal{O}$). Note that the inner/outer/mixed Voronoi edges are dual to the inner/outer/mixed Delaunay edges.

Based on the above definitions, the extraneous MA edges are the inner Voronoi edges whose both end points lie on the same curve segment. However, the dual of the main MA edges are the inner Voronoi edges (or its dual inner Delaunay edges) whose end points lie on two different curve segment. Therefore, the main idea of the proposed approach is to remove all the MA edges whose corresponding Delaunay vertices lie on the same boundary curve.

We start with labeling the sample points: Each segment of the shape is assigned a unique label; and all of its sample points are assigned the same label. The points that are common between two curve segments are called *junctions*, which are assigned a unique negative label to distinguish them from other sample points.

Filtering in our proposed method is not a pre- or post-processing step, but it is performed simultaneously with the MA extraction. To extract the MA, we use one-step crust and skeleton algorithm. Each Delaunay edge passes the *InCircle* test: If the determinant is negative and the corresponding Delaunay vertices have the same labels or one of them is a junction, that Delaunay edge is part of the crust. Otherwise, if the determinant is positive and the corresponding Delaunay vertices have different labels, its dual is added to the MA. Figure 10 illustrates the result of using our proposed algorithm to extract the MA of the river network presented in Fig. 6, in which the extraneous branches are automatically avoided.

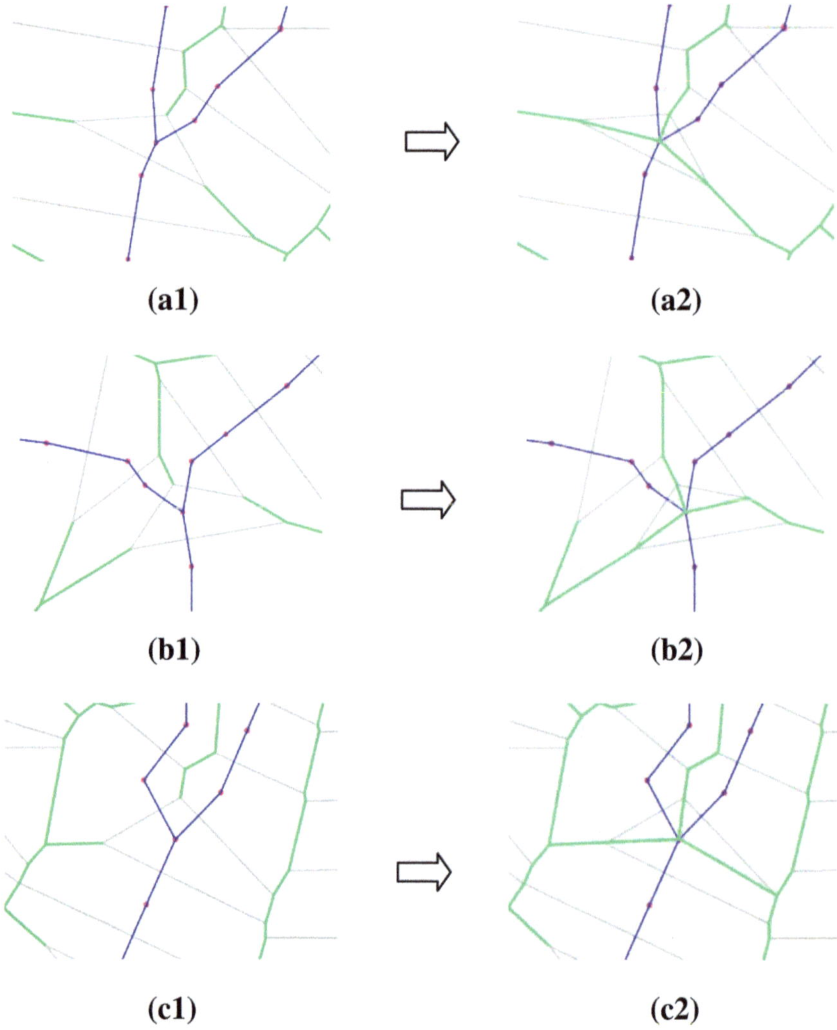

Fig. 11 Different configuration of the MA around the junction

5.1 Closing the Polygons

The calculated MA is a set of lines, whose union does not necessarily construct real polygons (catchments, here). A final check is needed to find and fix the gaps. For this, each junction is connected to three of the vertices of the Voronoi cell that contains the junction. Depending on the configuration of the MA around the junction, three cases may happen:

- The Voronoi cell of the junction point is a triangle (Fig. 11a1). Here, the three vertices of the Voronoi cell are connected to the junction point (Fig. 11a2).

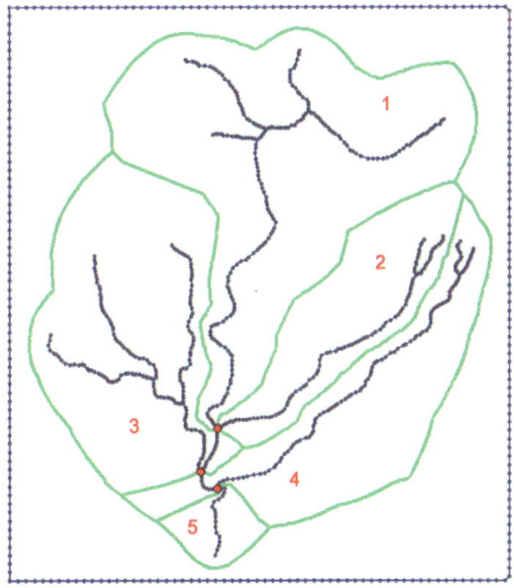

Fig. 12 Hierarchical delineation of catchments using the MA

- The Voronoi cell of the junction point has more than three vertices, and k of these vertices are dangle nodes. If $k = 3$ (Fig. 11b1), the three dangle vertices are connected to the junction point (Fig. 11b2). Otherwise, if $k < 3$ (Fig. 11c1), the k dangle nodes plus 3-k random vertices are connected to the junction (Fig. 11c2).

5.2 Hierarchical Delineation

Having extracted the MA using the one-step crust and skeleton algorithm, each branch of the streams is assigned a catchment polygon. Merging the polygons extracted for the branches of the same stream yields the catchment area of that stream. In our proposed method, however, a single polygon is already assigned to all the branches of the stream because they all have the same label (Fig. 12).

6 Implementation Results

We used the proposed approach for a case study and compared the result with a DEM-based method. Numerous techniques exist for automated extraction of catchment boundaries from DEM. The DEM is as a set of points in a Triangular Irregular Network (TIN) (Nelson et al. 1994; Jones et al. 1990), or a raster surface (Mower 1994). These methods usually simulate rainfall and assume that water always flows along the path of steepest descent.

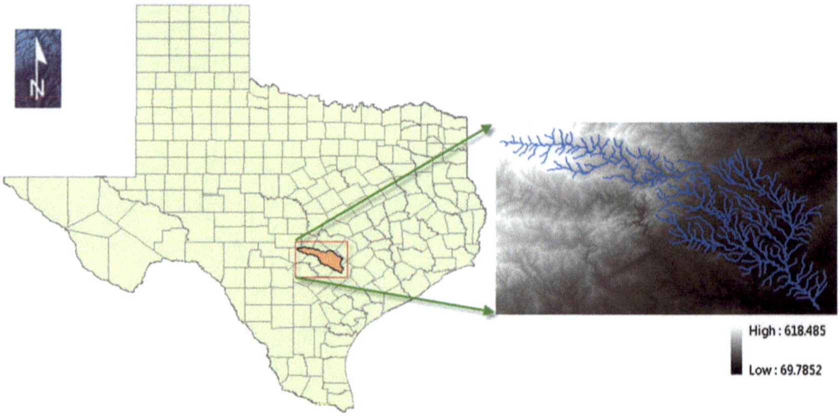

Fig. 13 Study area located at the upstream of San Marcos River in the south of Texas

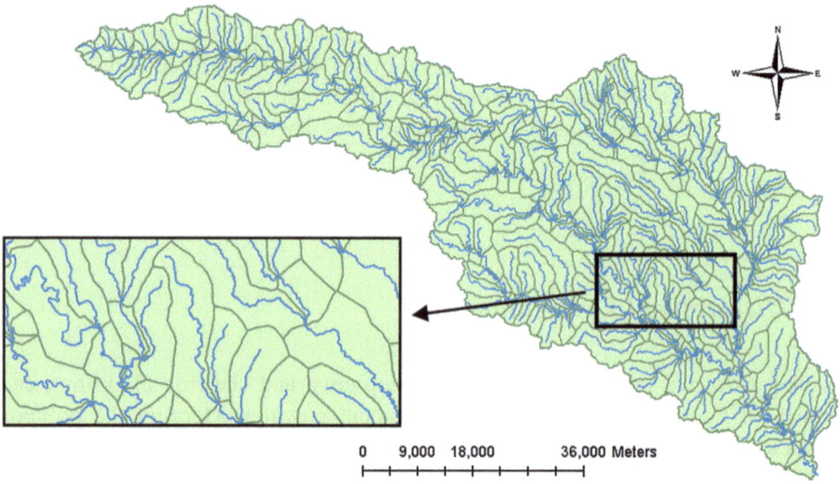

Fig. 14 Catchment area delineation of the study area from the river network using the proposed method

The study area is San Marcos basin located at the upstream of San Marcos River in the south of Texas State, USA (Fig. 13). It has an area of 3528 km^2 and its elevation varies from 70 to 618 m. The data used here consists of river data from the National Hydrography Dataset (NHD) and a raster DEM of the San Marcos Basin from the National Elevation Dataset (NED) with the spatial resolutions of 1 arc-second (approximately 30 m). Figures 14 and 15 respectively illustrate the catchment area delineated using the proposed approach and the DEM-based method implemented in ArcHydro extension of ArcGIS software, which are fairly similar.

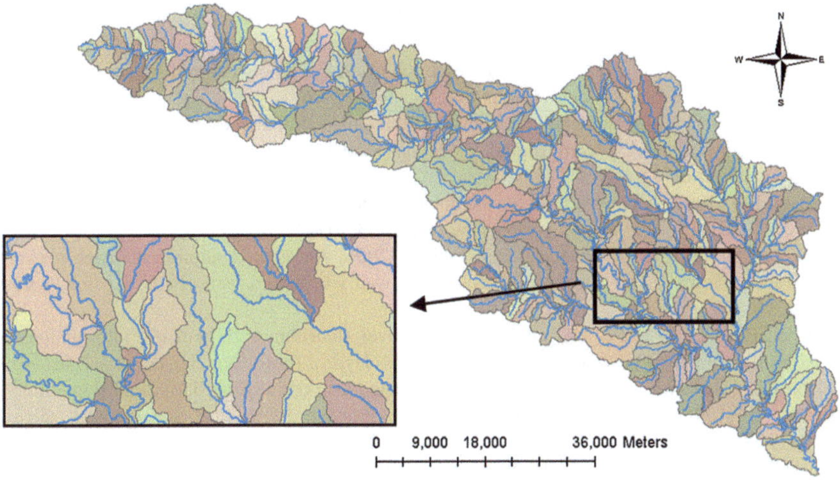

Fig. 15 Catchment area delineation of the study area from DEM using ArcHydro

7 Conclusion and Future Works

This chapter verifies the idea of Gold and Snoeyink (Gold and Snoeyink 2001) to use the MA of river networks as an approximation of the catchment areas. The advantages are: (1) It is simple and easy to implement (2) The method can handle very large areas and produce catchment polygons quickly. While, by increasing the spatial resolution of DEM or for a large area, the size of the raster DEMs required to delineate catchments would be large and the related processing like flow direction and flow accumulation can be so time consuming.

We also modify the algorithm proposed by Gold and Snoeyink to automatically avoid appearing extraneous branches in the MA. This solution is simple, easy to implement and robust to boundary perturbations of the sample points (river network).

The proposed method gives an approximation of the catchment area boundaries using the MA of the river networks. Although the output of the proposed method may not be as accurate as DEM-based methods, it can be efficiently used when DEM data is either inaccessible or in a poor quality.

References

Martz LW, Garbrecht J (1993) Automated extraction of drainage network and watershed data from digital elevation models. JAWRA 29:901–908

Turcotte R, Fortin JP, Rousseau A, Massicotte S, Villeneuve JP (2001) Determination of the drainage structure of a watershed using a digital elevation model and a digital river and lake network. J Hydrol 240:225–242

Chorowicz J, Ichoku C, Riazanoff S, Kim YJ, Cervelle B (1992) A combined algorithm for automated drainage network extraction. Water Resour Res 28:1293–1302

Martz LW, Garbrecht J (1992) Numerical definition of drainage network and subcatchment areas from digital elevation models. Comput Geosci 18:747–761

Mark DM (1984) Part 4: mathematical, algorithmic and data structure issues: automated detection of drainage networks from digital elevation models. cartographica. Int J Geogr Info Geovisualization 21:168–178

Tarboton DG (1997) A new method for the determination of flow directions and upslope areas in grid digital elevation models. Water Resour Res 33:309–319

Lin WT, Chou WC, Lin CY, Huang PH, Tsai JS (2006) Automated suitable drainage network extraction from digital elevation models in Taiwan's upstream watersheds. Hydrol Process 20:289–306

Yang W, Hou K, Yu F, Liu Z, Sun T (2010) A novel algorithm with heuristic information for extracting drainage networks from Raster DEMs. Hydrol Earth Syst Sci Discuss 7:441–459

Nelson EJ et al (1994) Algorithm for precise drainage-basin delineation. J Hydraul Eng 120:298

Mower JE (1994) Data-parallel procedures for drainage basin analysis. Comput Geosci 20:1365–1378

Jones NL, Wright SG et al (1990) Watershed delineation with triangle-based terrain models. J Hydraul Eng 116:1232

Li Z, Zhu Q, Gold C (2005) Digital terrain modeling: principles and methodology. CRC Press, USA

Gold C, Dakowicz M (2005) The crust and skeleton–applications in GIS. Second international symposium on Voronoi diagrams in science and engineering, pp 33–42

Dillabaugh C (2002) Drainage basin delineation from vector drainage networks. Joint international symposium on geospatial theory, processing and applications, Ottawa, Ontario, Canada

McAllister M (1999) The computational geometry of hydrology data in geographic information system. PhD thesis, University of British Columbia

Gold C, Snoeyink J (2001) A one-step crust and skeleton extraction algorithm. Algorithmica 30:144–163

Ledoux H (2006) Modelling three-dimensional fields in geo-science with the Voronoi diagram and its dual. PhD Thesis. School of Computing, University of Glamorgan, Pontypridd, Wales, UK

Karimipour F, Delavar MR, Frank AU (2010) A simplex-based approach to implement dimension independent spatial analyses. Comput Geosci 36:1123–1134

Blum H et al (1967) A transformation for extracting new descriptors of shape. Models for the perception of speech and visual form 19, 362–380

Amenta N, Bern MW, Eppstein D (1998) The crust and the beta-skeleton: combinatorial curve reconstruction. Graphical Models Image Process 60:125–135

Wenger R (2003) Shape and medial axis approximation from samples. PhD thesis. The Ohio State University

Siddiqi K, Bouix S, Tannenbaum A, Zucker SW (2002) Hamilton-Jacobi Skeletons. Int J Comput Vision 48:215–231

Attali D, Montanvert A (1996) Modeling noise for a better simplification of skeletons. In: IEEE international conference on image processing, vol 3. pp 13–16

Attali D, di Baja G, Thiel E (1995) Pruning discrete and semicontinuous skeletons. In: Proceedings of the 8th international conference on image analysis and processing, vol 974. pp 488–493

Chazal F, Lieutier A (2005) The Lambda Medial Axis. Graph Models 67:304–331

Attali D, Montanvert A (1994) Semicontinuous skeletons of 2D and 3D shapes. In: Proceedings of the second international workshop on visual form, pp 32–41

Giesen J, Miklos B, Pauly M, Wormser C (2009) The scale axis transform. In: Proceedings of the 25th annual symposium on computational geometry, pp 106–115

Karimipour F, Ghandehari M (2012) A stable Voronoi-based algorithm for medial axis extraction through labeling sample points. In: Proceedings of the 9th international symposium on Voronoi diagrams in science and engineering (ISVD 2012), New Jersey, USA
Ghandehari M, Karimipour F (2012) Voronoi-based curve reconstruction: issues and solutions. The international conference on computational science and its applications (ICCSA 2012), Lecture notes in computer science (LNCS), vol 7334. pp 194–207. Springer, Brazil
Giesen J, Miklos B, Pauly M (2007) Medial axis approximation of planar shapes from union of balls: a simpler and more robust algorithm. In: Proceedings of the 19th Canadian conference on computational geometry (CCCG), pp 105–108

Representing the Dual of Objects in a Four-Dimensional GIS

Ken Arroyo Ohori, Pawel Boguslawski and Hugo Ledoux

Abstract The concept of duality is used to understand and characterise how geographical objects are spatially related. It has been used extensively in 2D to qualify the boundaries between different types of terrain, and in 3D for navigation inside buildings, among others. In this chapter, we explore duality in four dimensions, in the context where space and other characteristics (e.g. time) are modelled as being in four dimensional space. We explain what duality in 4D entails, and we present two data structures that can be used to store the dual graph of a set of 4D objects. We also discuss applications where such data structures could be useful in the future.

1 Introduction

The concept of duality is used in geographical information systems (GIS) to understand and represent how things are connected, and to characterise spatial relationships. In two dimensions, one application is qualifying the spatial relationships between adjacent objects: as shown in Fig. 1, Gold (1991) uses two connected data structures to store simultaneously a polygonal map (where each polygon has certain attributes) and its dual (the boundaries between two map objects having certain attributes, e.g. the boundary type or the flow direction). He argues that the boundaries do not characterise *per se* any of the objects, but

K. Arroyo Ohori (✉) · H. Ledoux
Delft University of Technology, Delft, The Netherlands
e-mail: g.a.k.arroyoohori@tudelft.nl

H. Ledoux
e-mail: h.ledoux@tudelft.nl

P. Boguslawski
3D GIS Research Lab, Faculty of Geoinformation and Real Estate,
Universiti Teknologi Malaysia, Johor Bahru, Johor, Malaysia
e-mail: pawel@utm.my

A. Abdul Rahman et al. (eds.), *Developments in Multidimensional Spatial Data Models*, Lecture Notes in Geoinformation and Cartography,
DOI: 10.1007/978-3-642-36379-5_2, © Springer-Verlag Berlin Heidelberg 2013

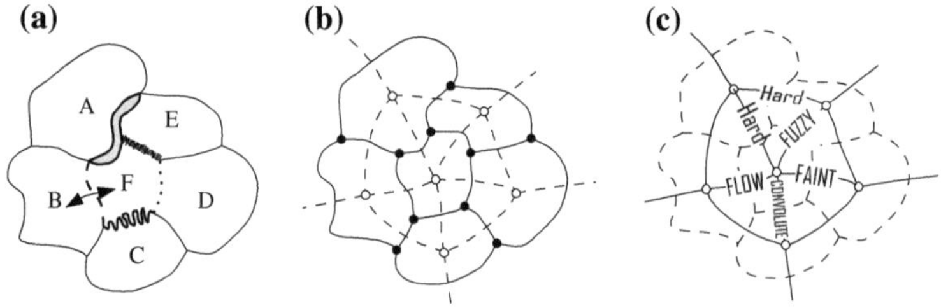

Fig. 1 **a** Six map objects and their boundaries. **b** The same map stored as a graph and its dual (*dotted lines*). **c** The dual graph is used to describe the relationships between adjacent polygons. [Figure after Gold (1991)]

rather the adjacency relationships that exist between them. The data structure used was the *quad-edge* structure of Guibas and Stolfi (1985). In three dimensions, duality also permits us to understand how different solids are spatially related (e.g. two rooms in a building are *adjacent*). Arguably the most known use of the dual is to model navigational paths inside three-dimensional buildings. Lee and Zlatanova (2008) and Lee and Kwan (2005) extract from a 3D building a graph that can be used in case of emergency, and Boguslawski et al. (2011) and Boguslawski (2011) perform the same using a data structure, the *dual half-edge* (DHE), which simultaneously represents the buildings (the rooms and their boundaries) and the navigation graph. With the DHE, the construction and manipulation operations update both representations at the same time, permitting the simultaneous modelling and characterisation of buildings. There are several other examples of duality in GIS: the Delaunay triangulation and the Voronoi diagram are often used to model continuous phenomena, these two structures being dual to each other. Dakowicz and Gold (2003) use them for terrain modelling, Lee and Gahegan (2002) for interactive analysis, and Ledoux and Gold (2008) for three-dimensional fields in geosciences.

In this chapter, we are interested in the concept of duality in four-dimensional space to model four dimensional objects. These objects are the result of the integration of a non-spatial dimension to the three dimensions of space, to create 4D objects where all dimensions are treated as spatial (Raper 2000). Examples of the non-spatial dimensions that can be used are: time (Peuquet 2002; Worboys 1994), scale (van Oosterom and Meijers 2011; Li 1994) and attributes (beyond 2.5D-type modelling). As van Oosterom and Stoter (2010) argue, the main advantage of such an integration is the *consistency* of data, both across space and the other dimensions modelled—with the proper validation functions, one can ensure that all the data for a given region is consistent across time or across different scales, for instance.

Adding an extra dimension implies that a 4D primitive has to be modelled: the *polychoron*[1], which is the 4D analogue of a polygon or polyhedron. To understand

[1] Also called a 4-polytope or a 4-polyhedron.

and characterise the spatial relationships between polychora (and between these and the lower-dimensionality primitives from which they are built), the dual graph of a set of polychora can be constructed and analysed. For example, in a 3D model of a building where rooms are represented by polychora we could locate people inside the building. With the dual graph, a user would be able to know where in 3D space a given user was at any time, when this person moved from a given room to another one, or the shortest path between any two rooms at any given time.

We first describe in Sect. 2 our terminology, the kinds of objects we are modelling in 4D, and the concept of duality. We then present in Sect. 3 potential data structures to store the dual graph of a partitioning of a 4D space. We first describe how *generalised maps* (Lienhardt 1994) can be used to extract and store the graph, and then we discuss how the DHE (Boguslawski et al. 2011; Boguslawski 2011) can be modified to store simultaneously both the partitioning and the dual in 4D. We analyse in Sect. 4 the storage of each structure, as one of our aims is to efficiently implement a 4D structure where the dual is available, and we briefly discuss in Sect. 5 potential applications of such data structures.

2 The Four-Dimensional Euclidean Space, and Duality

If we denote the three axes of the three spatial dimensions by x, y and z, then the axis w of the fourth dimension is perpendicular to that of all the spatial dimensions. In 4D Euclidean space (denoted $\mathbb{R}^4$), the simplest 4D primitive, called a *simplex*, is a 5-vertex polychoron and it is analogous to the triangle in 2D and to the tetrahedron in 3D. More generally, a d-dimensional simplex, also denoted as a d-simplex, is the convex hull of a set of $(d + 1)$ linearly-independent points in $\mathbb{R}^d$. Constructing a 4-simplex can be performed hierarchically: from a tetrahedron, we first embed its 4 vertices in 4D (with four coordinates each), then one new vertex is added, and finally 4 new edges must be constructed (these join the new vertex to the existing 4 of the tetrahedron). The resulting polychoron has 5 vertices, 10 edges, 10 triangular faces (2-simplices), and its boundary is formed by 5 tetrahedra (3-simplices).

By integrating 3D space and the extra dimension into 4D space, we ensure that there are no gaps or overlaps. This implies that we create a partitioning of $\mathbb{R}^4$. This can be achieved by keeping a so-called “universe polychoron” which encloses all the other polychora present in the model, in a manner similar to Liu and Snoeyink (2005). The structure we create is thus a partitioning into *cells*, where a 0-cell is a vertex, a 1-cell an edge, a 2-cell a polygon, a 3-cell a polyhedron, and a 4-cell a polychoron. Currently, no holes inside cells are allowed in our definition. We name a $(k - 1)$-cell incident to a k-cell a *facet* of it; a facet of a 4-cell is thus a 3-cell that lies in its boundary. This resulting partitioning forms a *cell complex* C, which is a finite set of cells having the following two conditions:

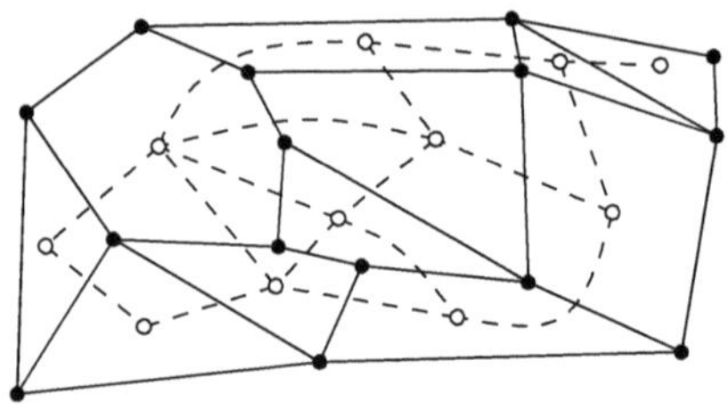

Fig. 2 A graph G (*solid lines*), and its dual graph G^* (*dashed lines*). For the sake of simplicity the dual edges to the edges on the boundary of G are not drawn

1. A facet of a k-cell in C is also in C;
2. The intersection of two cells σ_1 and σ_2 in C, denoted $\sigma_1 \cap \sigma_2$, is either empty or is in C.

A cell complex in $\mathbb{R}^4$ can be represented by a graph where the vertices and edges are embedded in $\mathbb{R}^4$ such that a set of vertices and edges implicitly represent a cell. Observe that since there are no holes allowed, the graph is connected.

Duality can have many different meanings in mathematics, but it always refers to the translation or mapping in a one-to-one fashion of concepts or structures. We use it here in the sense of the dual of a given graph. Let G be a planar graph (thus embedded in $\mathbb{R}^2$), as illustrated in Fig. 2 (solid edges); observe that G can also be seen as a cell complex in $\mathbb{R}^2$. The duality mapping is defined as follows; the dual graph G^* has a vertex for each face (polygon) in G, and these vertices are linked by an edge if and only if their two corresponding dual faces in G are adjacent (in Fig. 2, G^* is represented with dashed lines). Notice also that each polygon in G^* corresponds to a vertex in G, and that each edge of G^* (arcs in Fig. 2) is dual to an edge in G.

The concept of duality is valid in any dimension, as we consider a graph embedded in $\mathbb{R}^d$ as a d-dimensional cell complex. The mapping between the elements of a cell complex in $\mathbb{R}^d$ is simple: let C be a k-cell, the dual cell of C in $\mathbb{R}^d$ is denoted by C^* and is a $(d-k)$-cell. As a result, in four dimensions, a 0-cell becomes a 4-cell, and vice versa; a 1-cell becomes a 3-cell, and vice versa; and a 2-cell stays a 2-cell. Figure 3 shows the duality of a cell complex in $\mathbb{R}^3$.

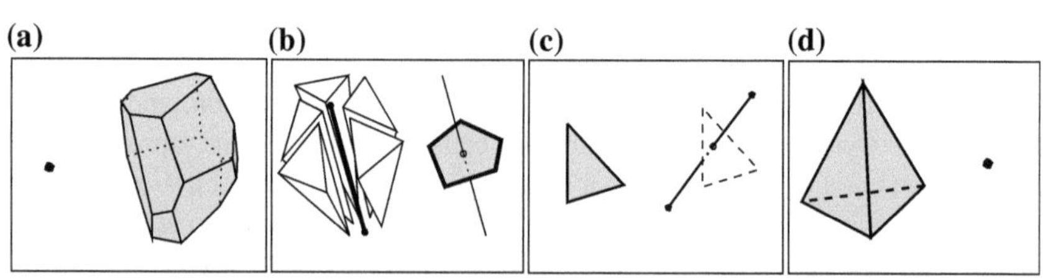

Fig. 3 Duality in a 3D cell complex

3 Potential Data Structures

There are several data structures that are able to represent models in four or more dimensions. Notable ones include: simplex-based ones (Paoluzzi et al. 1993; Shewchuk 2000), polytopal meshes (Sohanpanah 1989), (convex) decompositions of polytopes (Bulbul et al. 2009), and Nef polyhedra (Bieri and Nef 1988). However, despite the fact that they maintain various topological relationships, none of them provide efficient access to the dual graph of a model. We have nevertheless identified two candidate data structures that are able to do so, generalised maps and the dual half-edge. In this section, these are first introduced in their general form, and afterwards we specifically analyse how they could handle the dual graph in four dimensions.

3.1 Dual Half-Edge

The dual half-edge (DHE) structure, as proposed by Boguslawski et al. (Boguslawski et al. 2011, Boguslawski 2011), is a data structure that is able to represent a set of connected polyhedra forming a cell complex. It does so by simultaneously storing both the primal and the dual graphs of the objects, in a similar manner as the quad-edge structure of Guibas and Stolfi (1985) in 2D.

As shown in Fig. 4a, with the DHE each polyhedron is represented independently with an edge-based structure (a *b-rep* model), and adjacent polyhedra are linked together by their shared faces, which are represented by half-edges joining 3-cells. These form a graph of connections in the dual of the original (primal) graph. Both the primal and the dual graphs are identical in terms of structure (i.e. their basic elements and connections). Figure 4b shows an idea of the relationships that are stored for each edge. Since these graphs conform to the duality concept as explained in Sect. 2, the only cells that are needed to build a 3D model are the 0-cells (nodes) and 1-cells (edges); the nodes store the vertex coordinates, while the edges store the connections between the nodes. Meanwhile, the 2-cells (faces) and 3-cells (volumes) are only implicitly represented, but their attributes can be stored in their dual counterparts, the 1-cells and 0-cells in the dual graph.

However, an edge is not an atomic element in the DHE. Each edge consists of two half-edges, each of them being permanently connected with its corresponding half-edge in the dual. This pair, half-edge in the primal graph and half-edge in the dual one, is called the *dual half-edge*, and forms the atomic element in this model. Each half-edge is represented with five pointers which keep references to: an associated vertex, the next edge around a shared vertex, the next edge around a shared face, the second half-edge of the edge, and to the dual half-edge.

These five pointers are necessary to represent complex models including non-manifold cases—when two cells are only linked by a shared vertex or edge. However, the number of pointers can be reduced by one if only cells linked by a

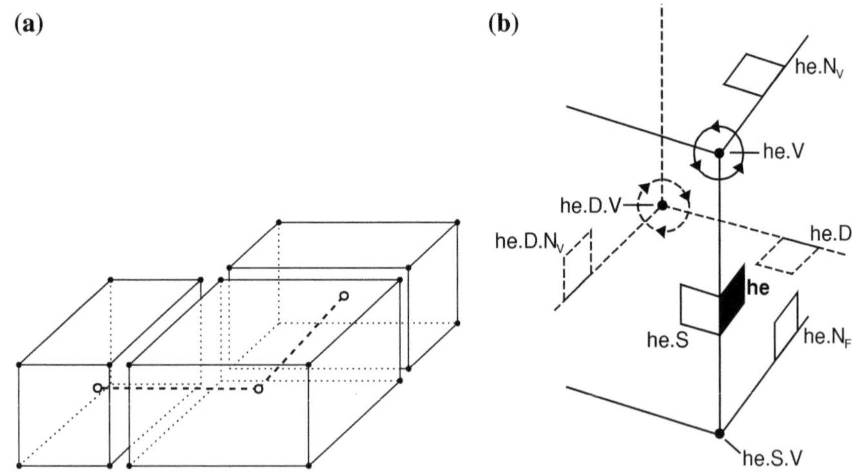

Fig. 4 The dual half-edge data structure in 3D. **a** The DHE models 3D subdivisions by representing the boundary of each polyhedron separately with a graph (*edges are solid lines*), and two adjacent polyhedra are linked together by the dual graph (*edges are dashed lines*). **b** The DHE pointer based data structure; the primal graph (*solid lines*) is connected permanently with the dual graph (*dashed lines*); *he* - original half-edge; S, N_V, N_F, D, V - pointers

shared face are taken into consideration. Additionally, a primal and dual half-edge pair can be merged and stored as a single record, since they are permanently connected—the number of pointers is reduced by one.

Using the data structure directly, without higher level construction operators would be extremely difficult—'manual' updating of pointers while edges are added to a model can easily cause many mistakes. Therefore, it is preferred to use the construction operators from Boguslawski (2011). They allow for model construction in an easy way, edge-by-edge, like in various CAD systems. Additionally, the dual graph is created automatically as the edges are added to the model and single cells are linked into a complex. These operators, used for modifications of the existing model, make only local changes in the primal and dual graph, and thus the whole dual graph does not need to be reconstructed after each modification.

During the construction process, the external cell, which encloses cells in a complex representing a modelled object, is automatically created. It can be considered as 'the rest of the world'. This external cell prevents topological inconsistencies at the boundary of the complex, where cells do not have an adjacent cell to connect to. Also, navigation can be implemented without testing if a boundary of the complex is approached.

Figure 5 shows one possible way to construct two cubes linked into one complex. It is based on CAD-like operators—Euler operators (Baumgart 1975; Braid et al. 1980; Mäntylä 1988) and extended Euler operators (Masuda 1993). First, two separate cubes are created (see Fig. 5a). Then, they are linked by a shared face (see Fig. 5b). It is possible to define different sequences which results in the same model. It should be noted that the external cell and dual graph are present at each step of the process, but for the sake of clarity the external cell and dual graph are not shown. The final model consists of three cells: two internal cubes and one external cell (see Fig. 5c).

The DHE was originally designed for 3D models. However, a single polychoron can be represented using the DHE without any modifications, except for the use of 4D coordinates. This is done by instead representing the polyhedra that lie on its boundary, in a similar manner as a 2D data structure is commonly used to represent a single polyhedron by storing the polygons in its boundary,

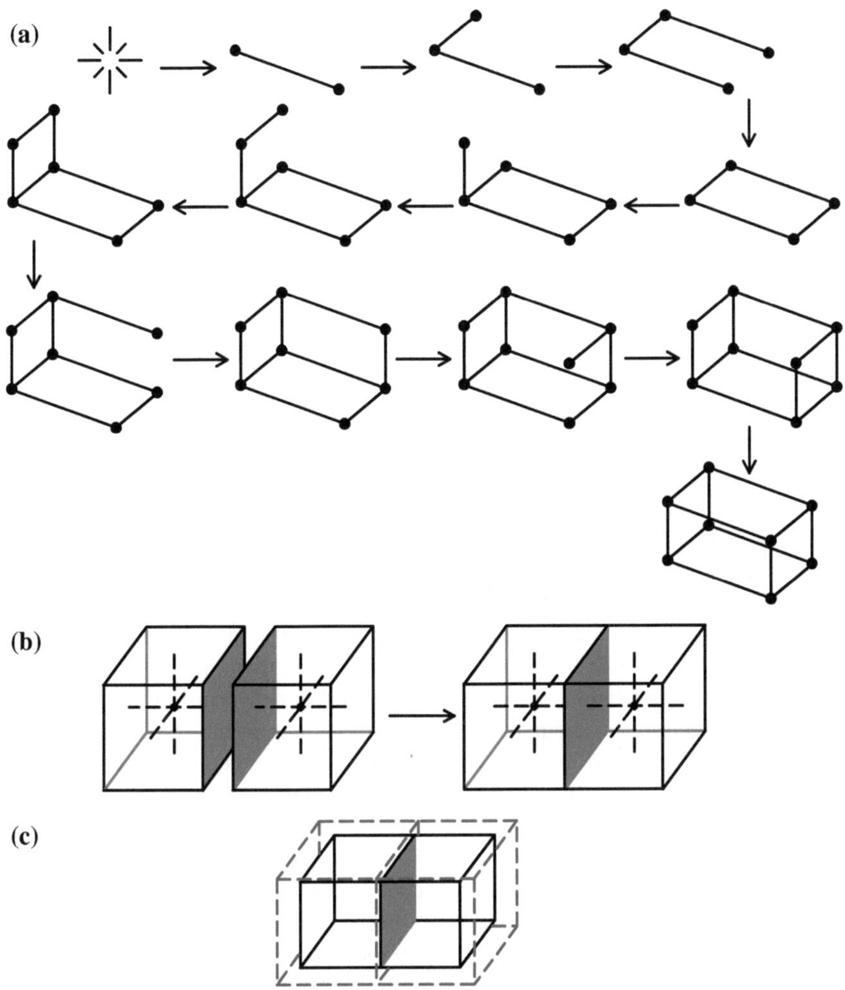

Fig. 5 Cell complex construction process. **a** A cube construction scenario. **b** Cubes share a common face (*grey*). Cells in the complex are connected using dual edges (*dotted lines*). **c** The resulting model consists of two internal (*solid lines*) and one external (*dotted grey lines*) cells

cf. Baumgart (1975). While this is not sufficient to represent a 4D cell complex with adjacent 4-cells or a non-manifold object, we believe that the data structure can be extended to represent objects in 4D. The biggest challenge is to correctly manage all the connections between the 4-cells, so as to fulfil the 4D duality rules.

3.2 Generalised Maps

Generalised maps (G-maps) are an ordered topological model developed by Lienhardt (1994) based on the concept of a combinatorial map, also known as a topological map, which was described by Edmonds (1960). They are roughly equivalent to the cell-tuple structure of Brisson (1989), but have been shown to be able to represent a wider class of objects known as cellular quasi-manifolds—manifolds that allow certain types of singularities, as long as every n-cell is incident to no more than two $(n + 1)$-cells.

Intuitively, a G-map is composed of two elements: a set of *darts*, each of which is defined by a unique combination of a specific n-cell from every dimension, and are often represented visually as half-edges or oriented edges; and *involutions* (α), bijective operators connecting darts that are related along a certain dimension. In this manner, α_0 joins vertices to form edges, α_1 connects consecutive edges within a face, α_2 connects adjacent faces within a volume, and so on.

One can obtain the connected darts that form a specific cell by the use of the *orbit* operator, which returns a (possibly ordered) set of darts that are reachable by following certain involutions only. To obtain the darts that are part of a certain i-cell only, one can start from any dart d belonging to the i-cell, following all involutions *except* for α_i. This is commonly denoted as $< \not\alpha_i > (d)$ (Lévy and Mallet 1999). Since α_i connects *adjacent* i-cells, not following it means remaining in the same cell. For simple construction, the *sew* operation is used, connecting two i-cells of the same dimension along the common part of their boundary. It does parallel traversals of two orbits, adding involutions that connect corresponding darts from each. Note that this implies certain ordering criteria in the orbit operator. Analogously, the *unsew* operation removes these involutions. An example of a 3D G-map representation of two adjacent cubes is shown in Fig. 6.

The aforementioned elements and operations represent the combinatorial structure of a generalised map. However, to represent the geometry of the model, an additional *embedding* structure is used. If only linear geometries are required, only the 0-dimensional point embeddings are actually needed. These store the coordinates of each vertex.

Since an α_i involution connects adjacent i-cells in the primal graph of a d-dimensional model, per definition α_{d-i} does so in the dual graph. These can therefore be easily swapped to convert a graph into its dual. For the 4D case, α_4 connects corresponding involutions for the dual of the nodes (0-cells), α_3 for the edges (1-cells), α_2 for the faces (2-cells), α_1 for the volumes (3-cells), and α_0 for the 4-cells.

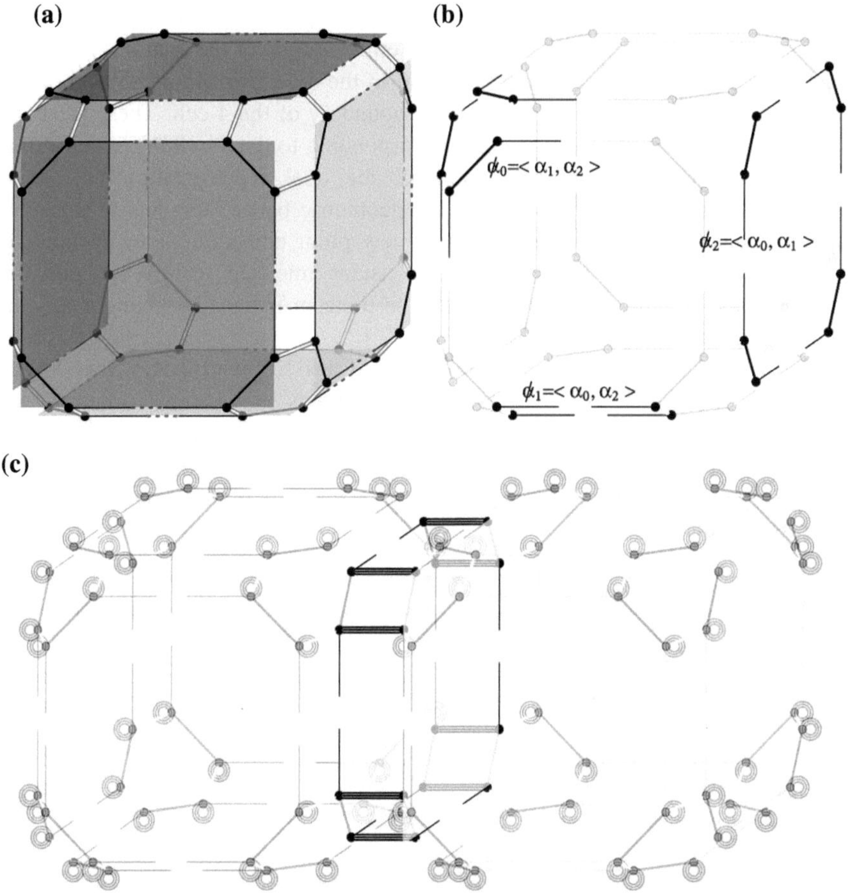

Fig. 6 A 3D G-map representation of a pair of adjacent cubes, showing the α_0 (*dashed*), α_1 (*solid*), α_2 (*double*), α_3 (*triple*), and $\not\alpha_i$ operators. **a** A G-map representation of a cube. **b** The $\not\alpha_i$ operator obtains all the darts belonging to a specific *i*-cell. Thus, $\not\alpha_0$ obtains the darts belonging to a vertex, $\not\alpha_1$ those belonging to an edge, and $\not\alpha_2$ those belonging to a face. **c** A G-map representation of two cubes. Note how the individual cubes have identical involutions to those of (**a**), with the addition of an α_3 involution that connects the two cubes at their common face. In the other darts, this involution is not used

Unlike the DHE that permits us to represent both the primal and the dual graph simultaneously, G-maps permits us to directly represent either one or the other. Transforming a 4D cell complex into its dual is however a straightforward operation, the combinatorial part of it being performed in linear time. Alternatively, any cell's dual can be directly obtained from the graph by interpreting an α_i involution as an α_{d-i} one. This is similar to how a Voronoi diagram is instead often manipulated from a Delaunay triangulation, cf. Boots (1974). Duality in the combinatorial structure of G-maps is therefore trivial to obtain, and can be done in real time.

Meanwhile, a geometric interpretation of the dual graph is also simple to get. Assuming linear geometries, only the point embeddings for the dual of the 4-cells need to be generated, e.g. using the centroid of the 4-cell, or simply an average of the point embeddings of the 0-cells in the boundary of the 4-cell. These are both easy to obtain using the $<\not{\alpha}_4>$ orbit corresponding to the 4-cell. Note that if the initial model (primal graph) is bounded, the dual representation will have unbounded cells. The running time of the geometric part of the duality transformation depends on the manner in which new point embeddings are computed. When each of these can be computed in constant time (e.g. using a few darts in their orbit), the entire transformation can be done in linear time. Otherwise, the complexity will be higher.

The process of the duality transformation of a 4D G-map is shown in Algorithm 1. For simplicity of explanation, three things are assumed to exist: an additional pointer to store the new point embeddings, a global list of embeddings, and a pointer from each embedding to a dart in its boundary. Note however that these are not strictly necessary. Their existence depends on the manner in which G-maps are implemented.

Algorithm 1: DUALTRANSFORM(G)

Input : *A 4D G-map* G
Output: The dual transformation of *G*
foreach *4-Cell c in G* **do**
    $p \leftarrow$ the centroid of c
    $O \leftarrow < \not{\alpha}_4 > (c.dart)$
    foreach *Dart d in O* **do**
        $d.embedding_{4\text{D}} \leftarrow p$
    end
end
foreach *Dart d in G* **do**
    swap($d.\alpha_0$, $d.\alpha_4$)
    swap($d.\alpha_1$, $d.\alpha_2$)
    swap($d.embedding_{0D}$, $d.embedding_{4D}$)
end

For consistency in our dual representation, we assume that there is an external (unbounded) cell. This ensures that applying the dual operation twice returns a model that is topologically equivalent to its original representation. The transformation of a 2D G-map into its dual is shown step by step in Fig. 7.

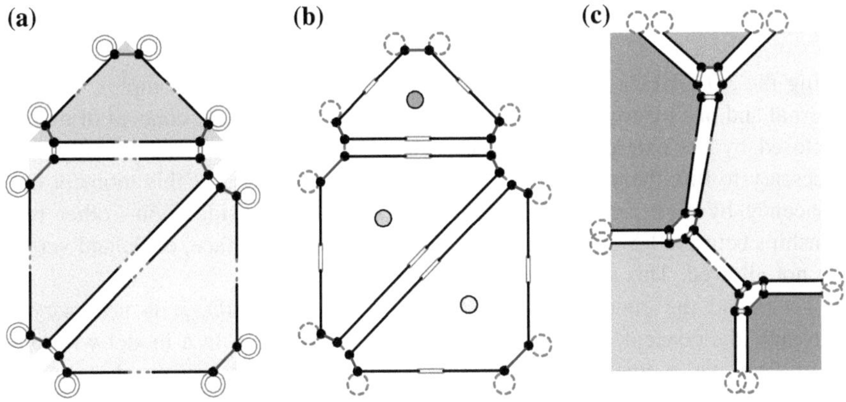

Fig. 7 A step by step transformation of a 2D G-map into its dual representation. **a** A 2D G-map representation of 3 adjacent triangles. **b** α_0 and α_2 are swapped, and a point embedding for each triangle is created. **c** The final result after linking to the new point embeddings

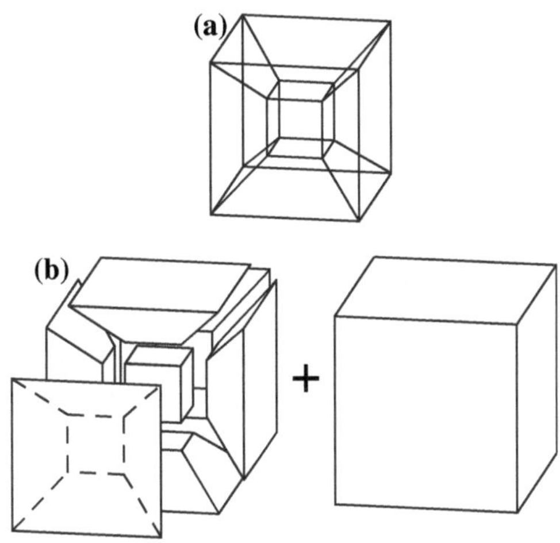

Fig. 8 A 3D projection of a tesseract: **a** a tesseract is a simple polychoron, **b** a tesseract is bounded by eight cubes, and can be represented as a cellular complex of these cubes. Note that they have different shapes due to the projection used

4 Storage of a 4D Cell Complex and its Dual

In this section we consider a 4D cell complex consisting of adjacent tesseracts and its dual. A tesseract, also known as a 4-cube or cubic prism, is the four dimensional analogue of a cube. As shown in Fig. 8, it is a closed four-dimensional polytope bounded by eight cubes. It contains 8 cubical 3-cells, 24 square 2-cells, 32 1-cells, and 16 0-cells.

4.1 Dual Half-Edge

Using the 3D DHE, a single tesseract can be represented as a complex of seven internal and one external 3-cells. This perfectly fits to the DHE concept of models enclosed by the external cell. Once each tesseract is created individually, it is necessary to link them into a 4D complex. Using the DHE, at this moment only adjacency by a shared 3-cell (a cube) is taken into consideration—other relationships between lower dimensional cells (i.e. by a shared face, edge, and vertex) are not allowed. This significantly simplifies the problem.

To extend the concepts of the original 3D DHE to 4D, it is necessary to introduce the concept of an external 4-cell into the model. In a model with only one object, e.g. a single tesseract, there will be two 4-cells—internal and external—connected into a complex by adjacent 3-cells. Thus, in the dual of this model, there would be two vertices, which correspond to the internal and external 4-cells. The vertex corresponding to the internal one can be calculated as the centroid of the tesseract; while the node for the external one may be located at infinity—these two dual nodes are connected by eight dual edges representing all the bounding 3-cells of the internal and external 4-cells. Technically, each 3-cell is represented by a bundle of dual edges, but since the bounding nodes of the edges in the bundle are geometrically the same, and the edges are connected in a radial cycle, they are considered as a single edge. Each cube in this example is represented by a bundle of 12 dual edges. Since every primal half-edge is associated with a dual half-edge, the number of dual edges is the same as the number of primal edges of the cell.

A simple calculation determines that the number of atomic elements, dual half-edges, required to represent the above model is 384: there are two tesseracts (internal and external) consisting of eight cubes each; each cube consisting of 12 edges; and each edge consisting of two DHEs. It should be noted that the 3D dual graph originally used to connect 3-cells of a complex is replaced by the new 4D graph connecting 4-cells. However, at the current stage of our research we cannot determine all the DHE connections between the dual edges, and the exact number of pointers necessary to represent these connections.

4.2 Generalised Maps

A complex of two or more adjacent tesseracts can be represented using a 4D generalised map (4-G-map), in which each dart has 5 involutions (α_0–α_4). The construction of the model proceeds incrementally, starting from the vertex level. A vertex is defined by a point embedding, where its 4D coordinates are stored. An edge is defined by creating a pair of darts, linked to each other along the α_0 involution. Four of these edges, linked along their shared vertices at the α_1 involution, form a square face. Six of these faces, linked along their shared edges at the α_2 involution, form a cube. Eight of these cubes, linked along their shared faces at the α_3 involution, form a tesseract. The resulting tesseract is thus formed of $2 \times 4 \times 6 \times 8 = 384$ darts and 16 point embeddings. Since each dart requires 6 pointers (one for each involution plus one for its point embedding), there are 2,304 pointers in the combinatorial structure.

The α_4 involutions have not been used up to this point. These are however used to link 3D-adjacent tesseracts together. Since no additional pointers are required, the total storage used for a cell complex of 4D tesseracts is the sum of the storage for each individual tesseract.

To obtain the dual of this model, the procedure described in Sect. 3.2 may be used. The α_i and α_{d-i} involutions are first swapped in the combinatorial structure. In this manner, α_0 becomes α_4 and vice versa, and α_1 becomes α_3 and vice versa. Afterwards, a new point embedding at the centre of each tesseract is created and linked to the darts on its boundary.

5 Discussion

We have shown how it is possible to store the dual graph of a 4D object by applying and extending existing data structures. G-maps already offer this possibility, although simultaneous storage of both graphs is not possible. The dual half-edge offers this possibility and is thus a promising alternative, especially as the dual graph is updated automatically while the primal is modified.

We also envision being able to use the 4D dual graph for various applications, navigation in 4D being an interesting possibility. For instance, it would make it possible to create a 3D indoor and outdoor way-finding application, where a user can select any given start and end points, and be given the best 3D route at any point in time, taking into account topological changes (e.g. a connecting corridor being only open during office hours).

We also plan to work on duality when holes/cavities are allowed in any dimension, up to 4D. An example of a 4D hole could be a section of a building being closed due to refurbishing work and thus inaccessible and removed from the graph. Note however that this assumption might not be true for all applications, e.g. emergency response. In a 3D representation, there would not be a natural connection between the building before and after the construction work, but it would be there in the 4D dual graph. This will allow us to fully utilise existing spatial datasets, and at the same time be able to represent a greater variety of situations.

Acknowledgments This research is supported by the Ministry of Higher Education in Malaysia (vote no. 02H97, Universiti Teknologi Malaysia) and by the Dutch Technology Foundation STW, which is part of the Netherlands Organisation for Scientific Research (NWO), and partly funded by the Ministry of Economic Affairs, Agriculture and Innovation. (Project code: 11300)

References

Baumgart BG (1975) A polyhedron representation for computer vision. In: Proceedings of the May 19–22, National computer conference and exposition, pp 589–596.

Bieri H, Nef W (1988) Elementary set operations with d-dimensional polyhedra. In: Computational geometry and its applications, Lecture notes in computer science, vol 333. Springer Berlin, pp 97–112.

Boguslawski P (2011) Modelling and analysing 3d building interiors with the dual half-edge data structure. PhD thesis, University of Glamorgan.

Boguslawski P, Gold CM, Ledoux H (2011) Modelling and analysing 3D buildings with a primal/dual data structure. ISPRS J Photogrammetry & Remote Sensing 66:188–197

Boots B (1974) Delaunay triangles, an alternative approach to point pattern analysis. In: Proceedings of association of American geographers 6:26–29.

Braid IC, Hillyard RC, Stroud IA (1980) Stepwise construction of polyhedra in geometric modelling. Brodlie KW (ed) Mathematical methods in computer graphics and design, Academic Press, In, pp 123–141

Brisson E (1989) Representing geometric structures in d dimensions: topology and order. In: Proceedings 5th annual symposium on computational geometry, ACM Press, Saarbrücken, West Germany, pp 218–227.

Bulbul R, Karimipour F, Frank AU (2009) A simplex based dimension independent approach for convex decomposition of nonconvex polytopes.In: Proceedings of geocomputation

Dakowicz M, Gold CM (2003) Extracting meaningful slopes from terrain contours. Int J Comput Geom Appl 13(4):339–357

Edmonds J (1960) A combinatorial representation of polyhedral surfaces. Notices of the American mathematical society 7.

Gold CM (1991) Problems with handling spatial data-the Voronoi approach. CISM J 45(1):65–80

Guibas LJ, Stolfi J (1985) Primitives for the manipulation of general subdivisions and the computation of Voronoi diagrams. ACM Trans Graph 4(2):74–123

Ledoux H, Gold CM (2008) Modelling three-dimensional geoscientific fields with the Voronoi diagram and its dual. Int J Geogr Info Sci 22(5):547–574

Lee I, Gahegan M (2002) Interactive analysis using Voronoi diagrams: Algorithms to support dynamic update from a generic triangle-based data structure. Trans in GIS 6(2):89–114

Lee J, Kwan MP (2005) A combinatorial data model for representing topological relations among 3D geographical features in micro-spatial environments. Int J Geogr Inf Sci 19(10):1039–1056

Lee J, Zlatanova S (2008) A 3D data model and topological analyses for emergency response in urban areas. In: Li J (ed) Zlatanova S Geospatial information technology for emergency response. Taylor and Francis, London, pp 143–168

Lévy B, Mallet JL (1999) Cellular modeling in arbitrary dimension using generalized maps. Tech rep, ISA-GOCAD

Li Z (1994) Reality in time-scale systems and cartographic representation. Cartographic J 31(1):50–51

Lienhardt P (1994) N-dimensional generalized combinatorial maps and cellular quasi-manifolds. Int J Computl Geom Appl 4(3):275–324

Liu Y, Snoeyink J (2005) The "far away point" for Delaunay diagram computation in $\mathbb{E}^d$. In: Proceedings 2nd international symposium on Voronoi diagrams in science and engineering, Seoul, Korea, pp 236–243.

Mäntylä M (1988) An introduction to solid modeling. Computer Science Press, New York

Masuda H (1993) Topological operators and Boolean operations for complex-based nonmanifold geometric models. Comput Aided Des 25(2):119–129

van Oosterom P, Meijers M (2011) Towards a true vario-scale structure supporting smooth-zoom. In: Proceedings of the 14th ICA/ISPRS workshop on generalisation and multiple representation, Paris.

van Oosterom P, Stoter J (2010) 5D data modelling: full Integration of 2D/3D space, time and scale dimensions, springer, In: Chap proceedings 6th international conference GIScience 2010, pp 311–324.

Paoluzzi A, Bernardini F, Cattani C, Ferrucci V (1993) Dimension-independent modeling with simplicial complexes. ACM Trans Graph 12(1):56–102

Peuquet DJ (2002) Representations of space and time. Guilford Press, New York

Raper J (2000) Multidimensional geographic information science. Taylor and Francis, London

Shewchuk JR (2000) Sweep algorithms for constructing higher-dimensional constrained delaunay triangulations. In: Proceedings of the 16th annual symposium on computational geometry, Hong Kong, pp 350–359.

Sohanpanah C (1989) Extension of a boundary representation technique for the description of n dimensional polytopes. Comput Graph 13(1):17–23

Worboys MF (1994) A unified model for spatial and temporal information. Comput J 37(1):26–34

Calibration and Accuracy Assessment of Leica ScanStation C10 Terrestrial Laser Scanner

Mohd Azwan Abbas, Halim Setan, Zulkepli Majid, Albert K. Chong, Khairulnizam M. Idris and Anuar Aspuri

Abstract Requirement of high accuracy data in surveying applications has made calibration procedure a standard routine for all surveying instruments. This is due to the assumption that all observed data are impaired with errors. Thus, this routine is also applicable to terrestrial laser scanner (TLS) to make it available for surveying purposes. There are two calibration approaches: (1) component, and (2) system calibration. With the intention to specifically identify the errors and accuracy of the Leica ScanStation C10 scanner, this study investigates component calibration. Three components of calibration were performed to identify the constant, scale error, accuracy of angular measurement and the effect of angular resolution for distance measurement. The first calibration has been processed using closed least square solutions and has yielded the values of constant (1.2 mm) and scale error (1.000008879). Using variance ratio test (F-Test), angles observation (horizontal and vertical) for Leica C10 scanner and Leica TM5100A theodolite have shown significance difference. This is because the accuracy of both sensors are not similar and these differences are 0.01 and 0.0075° for horizontal and vertical measurements, respectively. Investigation on the resolution setting for Leica C10 scanner has highlighted the drawback of the tilt-and-turn target. Using the highest resolution, Leica Cyclone software only able to recognize the tilt-and-turn target up to 10 m distance compare to 200 m for the black and white target.

Keywords Terrestrial laser scanner · Error source · Component calibration · Accuracy assessment

M. A. Abbas (✉) · H. Setan · Z. Majid · K. M. Idris · A. Aspuri
Universiti Teknologi Malaysia, Johor Bahru, Malaysia
e-mail: mohdazwanabbas@yahoo.com

A. K. Chong
University of Southern Queensland, Toowoomba, Australia

A. Abdul Rahman et al. (eds.), *Developments in Multidimensional Spatial Data Models*, Lecture Notes in Geoinformation and Cartography,
DOI: 10.1007/978-3-642-36379-5_3, © Springer-Verlag Berlin Heidelberg 2013

1 Introduction

The current terrestrial laser scanners (TLS) has considerably improved the measurement capability and the quality of the product in Geomatic field. Comparing with other measurement approaches, the processing involved in constructing a complete three-dimensional (3D) model is much simpler and faster (Varady and Martin 2002; Rabbani 2006; Pu 2008). Furthermore, current TLS can capture images using either an external or a built-in camera. This integration (TLS and camera) enables it to provide coloured point clouds and photorealistic 3D model.

However, similar to other instruments available in Geomatics field (total station, global positioning system, camera), observations from TLS is also impaired by errors (Reshetyuk 2009). In order to provide accurate data, standardized calibration routines exist for the traditional geodetic and photogrammetric instruments. According to Schofield and Breach (2007), calibration is the process of estimating the parameters that need to be applied to correct actual measurements to their true values. Since TLS is also used for Geomatics application, thus, calibration is also necessary in TLS measurement. Furthermore, the requirement of accurate data for surveying applications such as deformation, industrial survey and reverse engineering also has made this process crucial for TLS measurement. This is to ensure that TLS are suitable for various surveying applications.

2 Calibration of Terrestrial Laser Scanner

There are many instruments available to carry out surveying work and all of them require calibration in order to produce accurate data. This requirement also goes to TLS instrument, and it is a prerequisite for the extraction of precise and reliable 3D metric information from the point clouds (Reshetyuk 2009). Based on the error sources discussed in Sect. 3, there are several errors that can be evaluated during calibration. Two approaches available to investigate these errors, are either separately (component calibration) or simultaneously (system calibration) using statistical algorithms.

Schulz (2007) has differentiated between both calibration approaches, the system calibration derives either a mathematical model or a correction function, without knowledge of individual instrumental error and its influence. This is performed by properly design the network geometry of the control points. The component calibration is based on mainly knowledge-based modeling of the instrument and its instrumental errors. Individual error is investigated separately in a specific experimental setup.

This study implements component calibration in order to specifically identify the errors and accuracy of Leica ScanStation C10 scanner. There are three elements in the calibration and they are: (1) the determination of constant and scale errors, (2) accuracy of angular measurement and, (3) the effect of angular resolution for distance measurement.

3 Error Sources

As discussed by Böhler et al. (2003), Gordon et al. (2005), Schulz (2007), Lichti (2007) and Reshetyuk (2009), there are many error sources that may be modeled from TLS measurement. All of these error sources can be summarised into two groups as follows:

- Internal error which consists of errors from instrumental, data and resolution.
- External error that may include object related, environmental and georeferencing errors.

For this study, the Leica ScanStation C10 is evaluated as follows:

- Distance measurement system (constant and scale errors).
- Angle measurement system.
- Angular resolution.

3.1 Constant Error

Rueger (1996) defined that constant or zero error is caused by the virtual electro-optical origin or center of instrument is not being coincident with their physical center. This effect is a small correction which has to be added to all distance measurements. A number of factors may affect the value of the constant error and these factors may vary from instrument to instrument. Some instruments have constants which are signal strength dependent, while others are voltage dependent. The signal strength may be affected by the accuracy of the pointing or by prevailing atmospheric conditions. It is very important, therefore, that periodical calibrations are carried out.

3.2 Scale Error

Scale error is a scale factor in the measured distance. This error is largely due to the fact that the oscillator is temperature dependent. The quartz crystal oscillator ensures the frequency remains stable to within ± 5 ppm over an operational temperature range of –20 to 50 °C. The modulation frequency can, however, vary from its nominal value due to incorrect factory setting, ageing of the crystal and lack of temperature stabilization. Most modern short-range instruments have temperature-compensated crystal oscillators which have been shown to perform well. However, warm-up effects have been shown to vary from 1 to 5 ppm during the first hour of operation (Schofield and Breach 2007). Diode anomaly also causes scale error, as they can result in the emitted wavelength being different from its nominal value.

3.3 Accuracy of Angle Measurement System

According to Wunderlich (2003), the angular accuracy obtained by TLS is limited by the quality of the angular encoders. Hebert and Krotkov (1992) have found that poor synchronization also can be considered as sources of error in the angle measurement system. The beam deflection unit and distance measurement system should be accurately synchronised. Any deviation may result in errors along the scan line direction, for instance, skewed images. This type of errors is relatively constant as long as the scanner is stationary. Errors involved in angle measurement affect the calculation of 3D point coordinates.

3.4 Angular Resolution

Angular resolution can be described as the ability to detect small objects or object features in the point clouds (Böhler and Marbs 2005). Technically, there are two options offered by current TLS regarding this ability, the smallest possible increment angle between two successive points and the size of the laser spot itself to the object. Most scanners allow operator to manually choose the setup of this ability. Thoughts must be given to the selection of the setting because scanning time can be very expensive for choosing high resolution setting and vice versa for choosing low resolution where quality of the data is also relatively low.

4 Methodology

Due to similar measurement method between total station and TLS measurement, this study adapts calibration procedure of total station for TLS calibration. However, there are several differences as compared to total station calibration, such as target design, resolution setting and data reduction.

In total station calibration, the use of prism is required in order to obtain high accuracy measurement. In contrast, the use of prism or retro-reflectors are not recommended for TLS calibration since their high reflectivity (flaring and blooming) causes errors in the range measurement (Gordon et al. 2005). According to Lichti and Gordon (2004), flaring causes biases in range direction making the observed distance to the objects too short. And blooming causes targets to appear much larger than their actual size. To avoid those effects, this study employs the targets provided or recommended by the manufacturer. Five types of targets are used in this research and they are 76 mm (Fig. 1a) and 152 mm (Fig. 1b) tilt-and-turn (TT) targets. These targets are provided by manufacturer and additionally three types of black-and-white (BW) targets of various diameters of 150 mm (Fig. 1c), 200 mm (Fig. 1d) and 250 mm (Fig. 1e). For constant and scale errors determination, only the 76 mm TT

(a) (b) (c) (d) (e)

Fig. 1 Target used for the calibration, 76 mm (**a**), and 152 mm (**b**), tilt and turn target, black and white targets with diameter of 150 mm (**c**), 200 mm (**d**), and 250 mm (**e**)

target was used. For comparison with the high accuracy total station purposes, the second calibration employs 150 mm BW target. This is very important in order to ensure that the total station is able to measure the target. For the angular resolution calibration, all of the targets are used.

The main advantage of using TLS for data collection compared to total station is the data density. The capability of TLS to provide dense 3D point clouds make it applicable for various types of application especially 3D modeling. This density value can be set up by the user based on the angular resolution (low, medium or high). Even though, high resolution setting can give high accuracy data, the scanning process requires more time and large storage space to record the data. In contrast, choosing low resolution setting reduces the accuracy and this also affects the determination of target center.

Target pointing using total station can be determined directly by the user. The data reduction involves the calculation of horizontal distance from slope distance and vertical angle only. For the TLS, data reduction is rather complicated as prior to data reduction, the target center must be derived using the point clouds data. The center of the target is computed by finding the mean of the coordinates of all point clouds within the target.

By taking into account these differences between the total station and the TLS, this study should be carried out using three specific calibrations as follows:

- Constant and scale errors calibration using EDM baseline.
- Angular accuracy assessment.
- Angular resolution versus distance.

4.1 Constant and Scale Errors

For this experiment, EDM baseline site located at Permas Jaya, Johor Bahru was used. There are 10 pillars, which are maintained by the Department of Surveying and Mapping Malaysia (DSMM) and the pillars' position are determined using a high accuracy total station (Leica TCA2003) regularly.

Based on the technical specification of the scanner used for this study (Leica C10), this pulse based scanner is able to measure distance up to 300 m. However,

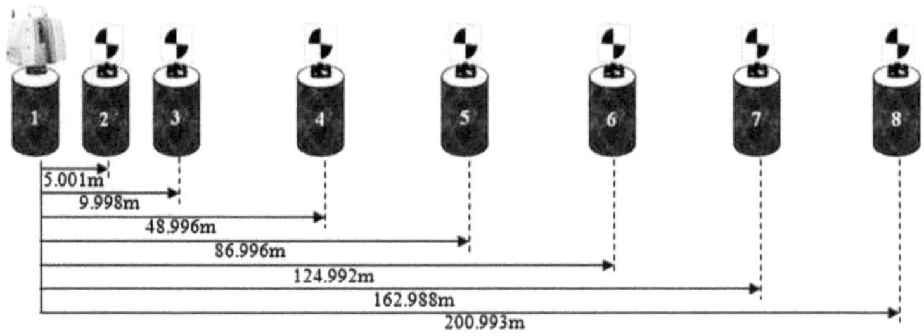

Fig. 2 TLS baseline calibration

based on the result of angular resolution calibration (Sect. 4.3), the targets can be recognized only up to 200.993 m (pillar 8). Therefore, the baseline calibration has been performed up to pillar 8 according to the following procedure (Rueger 1996):

- Setting up scanner at pillar 1, measurement was made to pillar 2, 3, 4 up to pillar 8 (Fig. 2).
- Move scanner to pillar 2 and proceed in a similar manner by measuring pillar 3, 4, 5 up to pillar 8.
- Finally, the scanner was moved to pillar 3 and targets were positioned at pillar 5, 6, 7 and 8.

As shown in Fig. 1, there are several targets being used in this experiment. However, in order to maintain the accuracy of scanning process, 6″ tilt and turn target provided by manufacturer were used. This is very crucial, for the purpose of reducing the constant error of the target.

Data obtained from the calibration were processed using linear regression solutions to resolve constant and scale errors as stated by Rueger (1996).

4.2 Angular Accuracy

This experiment compared the values of angles (horizontal and vertical) measured using high accuracy theodolite (Leica TM5100A) and scanner (Leica C10). In order to perform this test, eight targets were set up around the instruments as shown in Fig. 3.

As depicted in Fig. 4, a high accuracy theodolite (Leica TM5100A) with 0.5″ accuracy for both angular measurement (Leica 2002) was used for the benchmarking purposes. BW targets were chosen for this experiment due to its suitability for measuring the center of the target using theodolite while this could not be carried out on the TT target. To ensure the accuracy in the vertical angle measurement, all targets were set up at the same height.

The target scanning was carried out by using a medium resolution setting. This is based on the technique of Gordon et al. (2005). The authors noted in their study

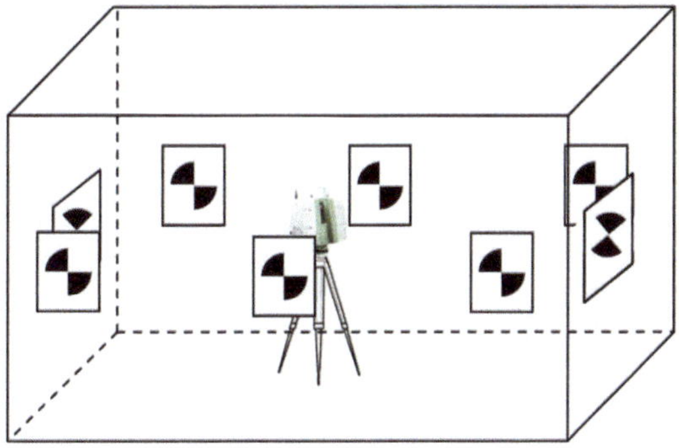

Fig. 3 Angular accuracy experiment

Fig. 4 High accuracy theodolite (Leica TM5100A)

that it is not essential to undertake high resolution scans due to the length of time required. Data captured from the scanning are set to a Cartesian coordinate system, and subsequently, with aid of Fig. 5 converted into spherical coordinate system using Eq. (1) (Reshetyuk 2009):

$$\begin{aligned} \text{Range}, (\text{r}) &= \sqrt{\text{X}^2 + \text{Y}^2 + \text{Z}^2} \\ \text{Horizontal\_direction}, (\varphi) &\tan^{-1}\left(\frac{\text{X}}{\text{Y}}\right) \\ \text{Vertical\_angle}, (\theta) &= \tan^{-1}\left(\frac{\text{Z}}{\sqrt{\text{X}^2 + \text{Y}^2}}\right) \end{aligned} \tag{1}$$

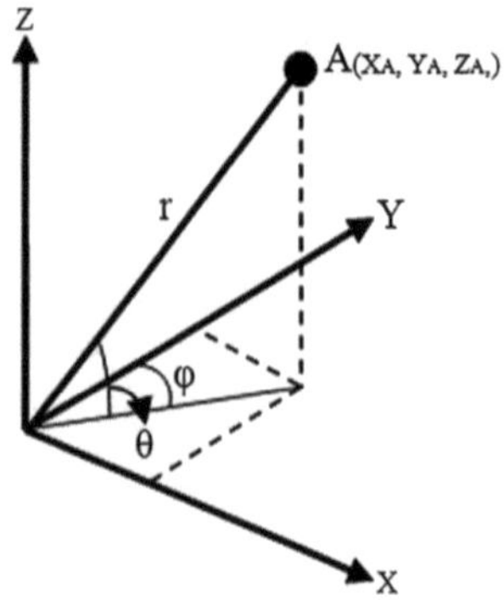

Fig. 5 Relationship between Cartesian and spherical coordinate system

Having the data in spherical coordinate system, the horizontal and vertical angles between the targets can be extracted. To provide adequate redundancy, the observations were repeated four times. For each set of observation, eight (8) horizontal and vertical angles were taken. With four repetitions, total number of observations were 64 and these are used to calculate a mean value and the precision of the mean value and these data were used to assess the accuracy of the angle measurement system.

As mentioned previously, data from theodolite is used as the benchmark to evaluate the data from Leica C10 scanner. This is carried out using the one-tailed F-variance ratio test for two population variances. The confidence level for the test is set as 95 % and the hypotheses of the test is:

H_0 The variances are not significantly difference.

H_A The variances are significantly difference.

The F-variance ratio test is carried out using Eq. (2) (Gopal 1999):

$$F = \frac{\sigma_2^2}{\sigma_1^2} \tag{2}$$

where, σ_1^2 refers to the variance for theodolite data and σ_2^2 refers to the variance for scanner data. The null hypothesis (H_0) is rejected if the calculated F value (Eq. 2) is higher than the critical F value (predicted from the F-distribution table) with selected level of significance (confidence level 95 % equal to 0.05 of significance level). The rejection of H_0 shows that the test parameters is not equal. In this study, the F-variance ratio test is used to analyse the similarity of the data from both sensors.

4.3 Angular Resolution

Once again, the baseline calibration pillar discussed in Sect. 3.1 are used for the angular resolution calibration (Fig. 6). In order to accurately evaluate the

Fig. 6 Angular resolution calibration using baseline technique

Table 1 Resolution setting for Leica ScanStation C10

Resolution setting	Horizontal spacing (m)	Vertical spacing (m)	Range (m)
Low	0.2	0.2	100
Medium	0.1	0.1	100
High	0.05	0.05	100
Highest	0.02	0.02	100

capability of the scanner to detect targets, several points have been established at baseline calibration to provide roughly 10 m interval between points up to a 100 m range. The five types of targets shown in Fig. 1 were measured in this calibration using four setting of resolutions (Table 1). To avoid the effect of incidence angle variation, all targets were aligned with respect to the scanner.

These resolution settings affect the determination of the targets center. Evaluation is carried out based on the capability of Cyclone software to automatically recognize the target. The result produces a yardstick for the resolution setting against distance. This can avoid operator from spending more time during scanning process to register all targets.

Table 2 TLS baseline calibration

Line	Measured value (m)	True values (m)
1–2	5.000	5.001
1–3	9.999	9.998
1–4	48.996	48.996
1–5	86.997	86.996
1–6	124.992	124.992
1–7	162.988	162.988
1–8	200.991	200.993
2–3	5.001	4.998
2–4	43.997	43.996
2–5	81.996	81.995
2–6	119.99	119.992
2–7	157.987	157.987
2–8	195.99	195.992
3–5	76.999	76.997
3–6	114.994	114.994
3–7	152.99	152.989
3–8	190.996	190.994
Standard deviation for observations		1.4 mm
Constant error		1.2 ± 0.6 mm
Scale factor		8.9 ± 5.1 ppm

5 Result and Analysis

As summarized in Table 2, constant and scale errors have been evaluated from the 17 observations. Utilising formula discussed in Sect. 3.1, both errors were calculated using true and measured values from the baseline calibration. The standard deviation for the measurement and both constant and scaling errors are presented.

For verification purpose regarding the significance of the calculated constant and scale errors, statistical analysis was performed. Known as *t* test, the analysis is carried out using Eq. (3) (Gopal 1999):

$$t = \frac{X}{\sigma_X} \tag{3}$$

where,

X = Parameter to be evaluate

σ_X = Standard deviation of parameter

The hypothesis have been set as follows:

H_0 The parameter is not significance to the scanner observation.

H_A The parameter is significance to the scanner observation.

The null hypothesis (H_0) is rejected if the calculated t value Eq. (3) is higher than the critical t value (predicted from the t-distribution table) with selected level of significance (confidence level 95 % equal to 0.05 of significance level). With the rejection of H_0, the test parameters is statistically significance (accept H_A). The results of the test are shown in Table 3.

Results of Table 3 show that null hypothesis was accepted for scale factor at 95 % confidence level. However, the hypothesis was rejected for constant error. A conclusion can be made that only constant error is significance to the scanner observation.

Calibration for the angular accuracy was carried out by scanning eight targets around the scanner. By using the data from theodolite as the reference, statistical tests were carried out to evaluate the similarity of the data yielded by Leica C10 scanner. Tables 4 and 5 summarise the results from F-variance ratio test for both angular measurement (horizontal and vertical angles).

With respect to Eq. (2), the null hypothesis was rejected where the difference between the two sensors is shown to be statistically significant with 95 % confidence level. The calculated value for F (for both Tables 4 and 5) exceeds the critical value of F (9.277). This difference may be due to the dissimilarity in accuracy between the sensors is too large. To investigate the cause of the difference, least square adjustments were used to calculate the residuals for the measurement of both sensors. These values are very appropriate for the identification of the accuracy of each sensor as depicted in Figs. 7 and 8.

Considering the residuals in terms of degrees in Fig. 7, it can be summarised that the accuracy for the horizontal encoder of the angle measurement system lies within ± 0.012 and $\pm 0.002^\circ$ for Leica C10 scanner and Leica TM5100A theodolite respectively. Consequently, the result of the difference in accuracy between both sensors is $\pm 0.01^\circ$ for horizontal angle measurement.

In Fig. 8, the distribution of the residuals indicates that the accuracy for the vertical encoder of the angle measurement system lies within ± 0.008 and $\pm 0.0005^\circ$ for Leica C10 scanner and Leica TM5100A theodolite. Thus, the difference in accuracy between both sensors is $\pm 0.0075^\circ$ for vertical angle measurement.

The differences in accuracy for both angular measurements has shown the cause of the result of the previous F-variance ratio tests. However, the angular accuracy produced from the residual pattern is quite different compared to the specifications

Table 3 Significance test for calibration parameters

Number of observation		17
Degree of freedom		15
Critical value for 't' (95 %)		1.753
Calibration parameters	Calculated 't'	95 %
Constant error	2.0	Yes
Scale factor	1.745	No

Note Yes—Significance, No—Not Significance

Table 4 F-variance ratio test for horizontal angles data

Observations	F-calculated	F-critical	Significance
Horizontal 01	35.909	9.277	Yes
Horizontal 02	783.818	9.277	Yes
Horizontal 03	402.613	9.277	Yes
Horizontal 04	75.602	9.277	Yes
Horizontal 05	125.783	9.277	Yes
Horizontal 06	57.803	9.277	Yes
Horizontal 07	125.311	9.277	Yes
Horizontal 08	35.954	9.277	Yes

Table 5 F-variance ratio test for vertical angles data

Observations	F-calculated	F-critical	Significance
Vertical 01	1998.088	9.277	Yes
Vertical 02	671.237	9.277	Yes
Vertical 03	354.524	9.277	Yes
Vertical 04	1278.218	9.277	Yes
Vertical 05	2438.917	9.277	Yes
Vertical 06	369.337	9.277	Yes
Vertical 07	2057.548	9.277	Yes
Vertical 08	690.570	9.277	Yes

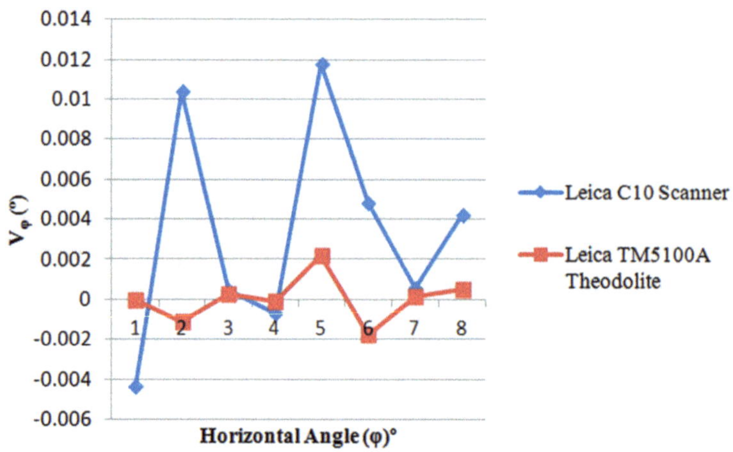

Fig. 7 Accuracy of the horizontal angle measurement for both sensors

of the manufacturer, which is 0.003° for both angular measurements. There is one possible cause that has reduced the accuracy of Leica C10. It may be the short distance used in this experiment (3.5–5.5 m) which has made unfavourable geometrical condition for the translation from linear values to angular values.

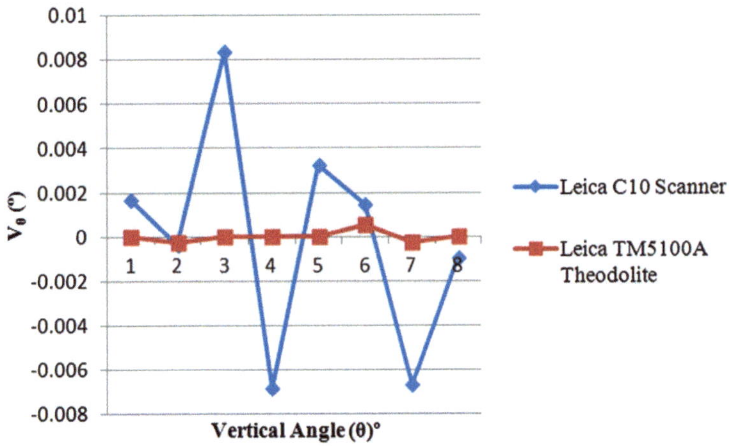

Fig. 8 Accuracy of the vertical angle measurement for both sensors

As discussed in Sect. 4.3, angular resolution setting can affect the ability of Cyclone software to recognize the target. With the intention to determine the resolution setting, calibration has been carried out using 5 types of target with 4 types of resolution setting. Using baseline calibration with additional station (refer to Sect. 4.3), Fig. 8 shows the results of this calibration.

Due to the additional scanning time requested for the highest resolution setting, up to an hour for each scan, this setting is frequently avoided by the user. Based on Fig. 9, by neglecting the result of highest resolution, the best range for the both TT targets is only 10 m. Conclusion can be made that those TT targets (76 and 152 mm) are only applicable for onsite target recognition as suggested by manufacturer. In contrast, BW targets are detectable by Cyclone software up to 50 m range. Medium resolution setting also applicable for BW targets for range within 20 m. However, a low resolution scan is not applicable for all types of targets, since the longest range it can achieve is only 10 m.

As mentioned in Sect. 4, both TT targets were provided by manufacturer and BW targets have been customized by the authors. Referring to the Fig. 9, conclusion can be made that the bigger size of target is easier for the Cyclone to recognize (based on TT targets). However, there is a little difference on the graph in which the 150 mm BW can be recognized further than 200 mm BW for high resolution setting. This occurs because of the margin used for 200 mm BW target is narrow (5 mm) compared to 150 mm BW target (refer Fig. 1). As depicted in Fig. 10a, the 150 mm BW target was recognized by Cyclone software. Figure 10b shows that Cyclone software was unable to recognize the target with small margin (200 mm BW).

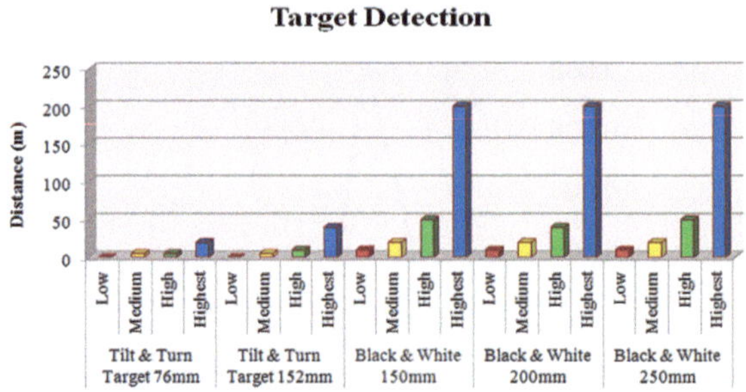

Fig. 9 Performance of the targets based on resolution setting against distance

6 Conclusion

Based on the data provided by TLS, this study developed the calibration techniques to determine and evaluate the accuracy of TLS measurement. This is very crucial due to the requirement of high accuracy in surveying applications. There are three different calibration techniques that were performed based on: (1) range, (2) angle measurements and (3) resolution setting. Calibration of range measurement has identified the values of constant and scale error. With the aid of t test, those systematic errors have been statistically confirmed as significant for the calibrated scanner. Using Leica TM5100A theodolite as a benchmark, angular accuracy of the Leica C10 scanner has been assessed. F-variance ratio test indicated that angular measurement (horizontal and vertical) for both sensors are significantly different. This is due to the difference in accuracy as illustrated by residuals pattern for both measurements. To avoid a cumbersome and time consuming procedure to register all the targets during scanning process as suggested by manufacturer, angular resolution calibration has been carried out to determine the relationship between resolution and target type. Results show that both TT targets are not suitable for target recognition during processing. In contrast, BW targets are better because it can be detected up to 20 m distance for medium resolution setting. For longer distance, high and highest resolution settings are suitable.

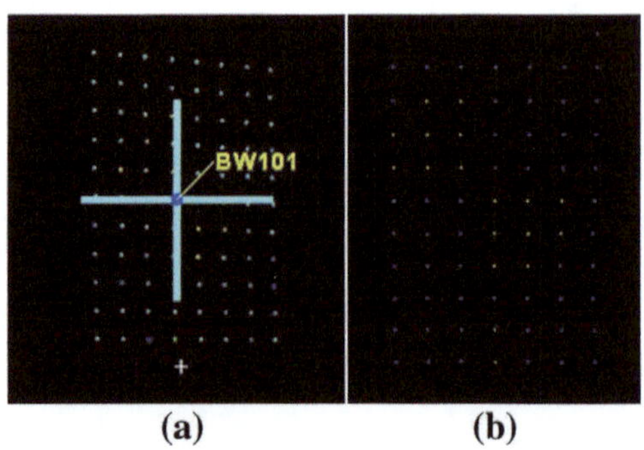

Fig. 10 Target recognition by Cyclone software for 150 mm BW (**a**), and 200 mm BW (**b**) targets

References

Böhler W, Bordas VM, Marbs A (2003) Investigating laser scanner accuracy. The international archives of the photogrammetry, remote sensing and spatial information sciences, vol XXXIV(Part 5/C 15):696–701

Böhler W, Marbs A (2005) Investigating laser scanner accuracy. Institute for spatial information and surveying technology. FH Mainz, University of Applied Sciences, Mainz, Germany

Gopal KK (1999) 100 statistical test. SAGE Publications Ltd, Thousand Oaks

Gordon S, Davies N, Keighley D, Lichti D, Franke J (2005) A rigorous rangefinder calibration method for terrestrial laser scanners. J Spat Sci 50(2):91–96

Hebert M, Krotkov E (1992) 3D measurements from imaging laser radars: how good are they? Image Vis Comput 10(3):170–178

Leica (2002) TPS1100 profesional series-technical data. Leica Geosystems AG, Heerburgg

Lichti DD, Gordon SJ (2004) Error propagation in directly georeferenced terrestrial laser scanner point clouds for cultural heritage recording. In Proceeding of FIG Working Week, Athens

Lichti DD (2007) Error modelling, calibration and analysis of an AM-CW terrestrial laser scanner system. ISPRS J Photogram Remote Sens 61(2007):307–324

Rabbani T (2006) Automatic reconstruction of industrial installations using point clouds and images. A thesis for the degree of Doctor of Philosophy at TU Delft

Reshetyuk Y (2009) Self-calibration and direct georeferencing in terrestrial laser scanning. Doctoral thesis in infrastructure, Royal Institute of Technology (KTH), Stockholm

Rueger JM (1996) Electronic distance measurement: an introduction, 4th edn. Springer, Berlin

Schulz T (2007) Calibration of terrestrial laser scanner for engineering geodesy. A dissertation submitted for the degree of doctor of scinces, Technical University of Berlin, Berlin

Schofield W, Breach M (2007) Engineering surveying, 6th edn. Published by Elsevier Ltd, Amsterdam

Pu S (2008) Automatic building modeling from terrestrial laser scanning. Springer, Berlin

Varady T, Martin R (2002) Reverse engineering. Chapter in handbook of computer aided geometric design, Elsevier Science B. V

Wunderlich TA (2003) Terrestrial laser scanners-an important step towards construction information. In Proceeding of FIG Working Week, Paris

A Methodology for Processing Raw LiDAR Data to Support Urban Flood Modelling Framework: Case Study—Kuala Lumpur Malaysia

Ahmad Fikri Abdullah, Zoran Vojinovic and Alias Abdul Rahman

Abstract High quality representation of the topographic and the correct representation of significant urban features would be a fundamental foundation to a better urban flood model. Without such a representation, simulation of flood behaviours would be less successful as the flow patterns were completely dependent on ground levels and the shape of the features. Typically, such data can be obtained via Light Detection and Ranging (LiDAR) surveys. The process of turning raw LiDAR data into a useful Digital Terrain Model (DTM) involves careful processing and application of thinning, filtering and interpolation algorithms. Filtering is a process of automatic detection and interpretation of bare earth and objects from the point cloud of LiDAR data, which results in the generation of a DTM. To date, many filtering algorithms have been developed, and in a more general sense, many of them have become standard industry practice. However, when it comes to the use of a DTM for urban flood modelling applications, these algorithms cannot be always considered suitable. Depending on the terrain characteristics, they can even lead to misleading results and degrade the predictive capability of the modelling technique. This is largely due to the fact that urban environments often contain a variety of features (or objects) such as buildings, elevated roads, bridges, curbs and others which have the ability to store or divert flows during flood events. As these objects dominate urban surfaces, appropriate filtering methods need to be applied in order to identify such objects and to represent them correctly within a DTM so that the DTM can be used more safely

A. F. Abdullah (✉)
Department of Biological and Agricultural Engineering, Universiti Putra Malaysia, 43400 Serdang, Selangor, Malaysia
e-mail: fikri@eng.upm.edu.my

Z. Vojinovic
Department of Hydroinformatics and Knowledge Management, UNESCO-IHE, Westvest 7 2611 AX, Delft, Netherlands

A. Abdul Rahman
Department of Geoinformatics, Universiti Teknologi Malaysia, UTM Skudai 81310 Johor, Malaysia

A. Abdul Rahman et al. (eds.), *Developments in Multidimensional Spatial Data Models*, Lecture Notes in Geoinformation and Cartography,
DOI: 10.1007/978-3-642-36379-5_4, © Springer-Verlag Berlin Heidelberg 2013

in modelling applications. The work described in this chapter concerns improvements of a LiDAR filtering algorithm. The key characteristics of this improved algorithm are: ability to recover curbs and the use of appropriated roughness coefficient of Manning's value to represent close-to-earth vegetation (e.g. grass and small bush). The results of the improved algorithm were demonstrated using Kuala Lumpur (Malaysia) as a case study. Improvement, in terms of a difference in flood depths and flood flows were observed between the hydraulics models built from several available filtering algorithms and the improved algorithm (MPMA). The overall results suggest that the improvement made in MPMA can lead to some difference in model results, which may in some cases be significant with a tendency towards incorrect flood flow by those models in which such features are not properly represented.

Keywords Urban flood modelling · LiDAR filtering algorithms · Digital terrain models (DTMs) · Urban features

1 Introduction

Urban Flooding has become an important issue in urban drainage planning and management. Proper drainage systems and analysis of their hydraulic performance to improve flood mitigation require accurate simulation models that applied couple drainage modelling based on detailed system data. Current modeling and simulation techniques are based on a very detailed computer representation of the underground sewer network which allows accurate sewer flow modeling by dynamic flow routing. In a mean time, the surface data to describe relevant surfaces are still very coarse. It has been studied and recognized that surface flow processes heavily affect the flooding both in terms of flood depth and extend. Considerable attention has been given to the acquisition of good geometric and topographical data at adequate resolution in order to describe the primary features of the flow paths through the urban area. In this respect, a Digital Terrain Model (DTM) represents one of the most essential sources of information that is required by flood managers. A DTM refers to a topographical map, which contains terrain elevations, and as such, it is used to characterise the terrain (or land) surface and its properties. It is a representation of the Earth's surface (or subset of it) and it strictly excludes features such as vegetation, buildings, bridges, etc.

The concept of urban flooding through minor system, i.e., underground pipes and major system, i.e., urban surfaces are used in a controlled manner. It is commonly referred to as coupling simulation model. Typically, urban flood modelling practice concerns the use of 1D, 1D/1D, 2D and 1D/2D modelling approaches; see, for example, Chen et al. (2005), Garcia-Navarro and Brufau (2006), Hunter et al. (2007, 2008), Kuiry et al. (2010), Price and Vojinovic (2011). Mark et al. (2004) demonstrated how the 1D modelling approach can be used to incorporate interactions between (1) the buried pipe system, (2) the streets (with open channel flow) and (3) the areas flooded with stagnant water. Djordjevic et al. (2005) have implemented a dual drainage concept (which represents a combination of minor and major systems) within a 1D model. Vojinovic and Tutulic (2009) have explored the difference in predictive capabilities of 1D and 1D/2D modeling approaches for the purpose of urban flood analysis across irregular terrains and their corresponding damage estimation. Also, Vojinovic et al. (2011) have shown how different terrain data resolutions, features such as roads and building structures, and different friction coefficients can affect the simulation results of 1D/2D models.

In conjunction with the advances in simulation techniques and the developments in data acquisition and GIS technologies have enabled explicit and more consistent treatment of coupling system interaction. Light Detection and Ranging (LiDAR) surveys enable the capture of spot heights at a spacing of 0.5–5 m with a horizontal accuracy of 0.3 m and a vertical accuracy of 0.15 m. Most of LiDAR surveys result in a substantial amount of data, which requires careful processing before it can be used for any application. This leads not only to more technically design of both sub-systems, but also to more reliable simulation of urban flooding and analysis of the consequences (Schmitt et al. 2002). The input DTM for flood modelling often needs to be at small resolution and re-sampling LiDAR data could result in loss of important terrain information that relate to the presence of streets and curbs. The desired target for DTM generation is bare earth points. Therefore, it is crucial to filter or extract bare earth points from LiDAR data. Thinning, filtering and interpolation are techniques that are used in the processing of LiDAR data. Various filter methods have been developed to classify or separate raw LiDAR data into ground and non-ground data. However, none of automated filter processes is 100 % accurate so far (Romano 2004). Manual editing of the filtering results is still needed (Chen 2007). Efforts are still needed to improve the performance of filter algorithms. Airborne LiDAR technology is still developing rapidly in both sensor and data processing. The competition between LiDAR sensor manufactures is mostly focused on increasing laser pulse repetition rates to collect more data points. The pulse repetition rate has increased from less than 50 kH in 2001 (Flood 2001) to 250 kH now (Lemmens 2007). High-density data make it possible to represent terrain in much detail. However, high-density data lead to a significant increase in the data volume, imposing challenges with respect to data storage, processing and manipulation (Sangster 2002). Although LiDAR data have become more affordable for users due to the gradual reduction of the costs of LiDAR data collection, how to process the raw LiDAR data effectively and extract useful information remains a big challenge (Chen 2007). Furthermore, because of the specific characteristics of LiDAR data, issues such as the choices of modelling methods, interpolation algorithm, grid size, and data reduction are challenging study topics for the generation of a high quality DTM from LiDAR data. The above described works are mainly related to LiDAR data processing in an attempt to make DTM or Digital Surface Model, DSM as accurate as possible. There have been some works on the application of LiDAR data in the context of flood modelling. Very few of them have discussed applicability of LiDAR data to support flood modelling in a complex urban area. Schmitt et al. (2004) introduced the concept of urban surface detailing in a small area and connected it to urban drainage modelling. Norman (2007) described detailing of complex urban areas including subtle surface distinction such as street, curbs and sidewalls. Haile and Rientjes (2005) described the sensitivity of LiDAR DTM resolution on urban flood modelling. An important challenge in using LiDAR data processing for urban flood modelling is to define the bare earth elevation that is used in the flood model to simulate surface elevation gradients between model grid elements. This is largely due to the fact that urban environments often contain a variety of features (or objects), which have functions of storing or diverting flows during flood events. As these objects dominate urban surfaces appropriate filtering methods need to be applied in order to identify such objects and to represent them correctly within a DTM so it can be used more safely in modelling applications. However, most of the current filtering algorithms are designed to detect tall vegetation and free-standing buildings only and features such as roads, curbs, elevated roads, bridges, rivers, river banks and close-to-earth vegetation are always difficult to detect. Therefore, further improvements of LiDAR filtering techniques are needed so that modelling efforts can generate more fruitful results.

2 Modification of the Progressive Morphological Algorithm (MPMA3)

The work concerning development of MPMA2 algorithm has been discussed in detail in Abdullah et al. 2011a, b. In the present work, the MPMA3 algorithm has been further improved by focusing on recovering of curbs and representation of closed-to-earth vegetation as roughness coefficient which has led to a new algorithm MPMA3. The overall flowchart for this new filtering algorithm is shown in Fig. 1. The overall aspects of MPMA3 algorithm are:

1. Ability to detect curbs using data fusion and vectorisation processes.
2. Ability to recover the curbs using the sub-grid concept.
3. Ability to detect closed-to-earth vegetation, to remove them from a DTM and to incorporate with the appropriate roughness coefficient (Manning's value).

2.1 Assumptions for Curbs Detection

Curb detection on the urban DTM is an important and necessary task for the preparation of a detailed urban surface as an input for urban flood model. Usually height of the curbs range from 5 to 10 cm. At the same time LiDAR data has vertical accuracy of 5–15 cm. It is a challenging task to determine the curb height

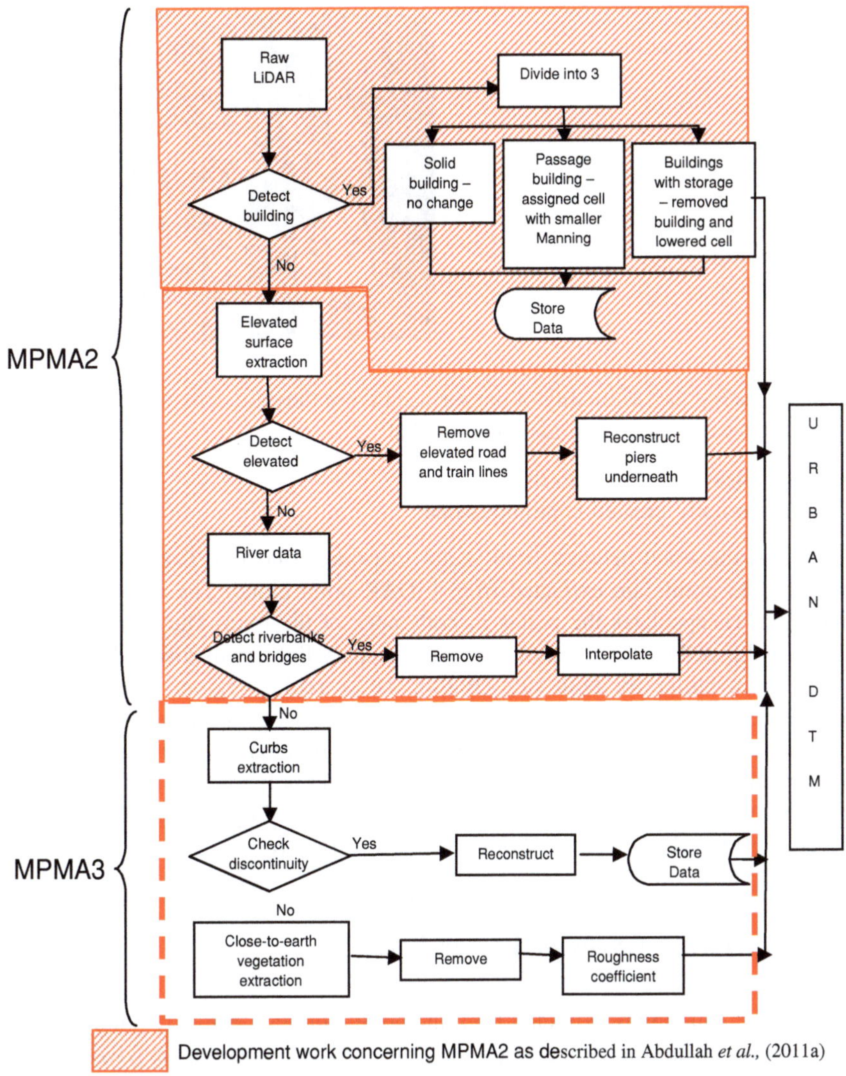

Fig. 1 Flowchart for the development of MPMA 2 and MPMA 3 algorithms

accurately because in many places the curb height is too small or there is no curb at all. The curb height is important in simulation model because of the location of street inlets and the possible entering of the water from the street to the private ground. Street inlets are located at the foot of curb wall. In case the street inlets are located on the top of curb wall by mistake, flood modelling is going to be affected.

In this chapter a semiautomatic approach for detection of road has been applied. This approach combines the vector data and raw LiDAR point cloud without converting it into any other raster format. The idea behind this approach is that raw

Fig. 2 The searching regions developed from vector maps

points preserve the originality of the surface before they are converted to any other formats. It works based on assumptions that:

1. Curbs lies very closed to the bare earth.
2. Curbs usually created continuous line.

At first the street network from vector maps are collected. Taking this as seed point, curbs points are extracted from LiDAR point cloud. The curbs extraction algorithm encapsulates the points in both sides of the road. The point searching region is limited by the input of the road boundary lines from road polygon maps with 30 cm buffer for both side. Within the searching region, points with the difference of 5–10 cm from the average road height will be selected as candidates for curbs line. The road average height is provided as input to this algorithm. The curbs extraction algorithm is guided by points on the road polygon and will search next curbs points and subsequently extracting points bounded by that searching region. Figure 2 shows the searching regions developed from vector maps.

2.2 Converting Curbs to Vector Form

Curbs candidate points extracted from the point cloud are converted into grid and then converted into a vector map (see Fig. 3). The converting into vector map process takes into account the orientation of groups of cells and fits lines to them. This is based on the sweep line Voronoi algorithm devised by Steven Fortune along with some filtering techniques (Fortune 1985). All cells that have a neighbouring cell with a value sufficiently different to it are turned into vectors.

2.3 Recovering of Curbs Using Sub-Grid

High resolution grid are often needed to divide the data set into small tiles geographically. This may result in poor performance when dealing with large study area. To address this issue, a sub-grid was adopted to create a finer set grid

Fig. 3 The example of curbs candidate in vector form

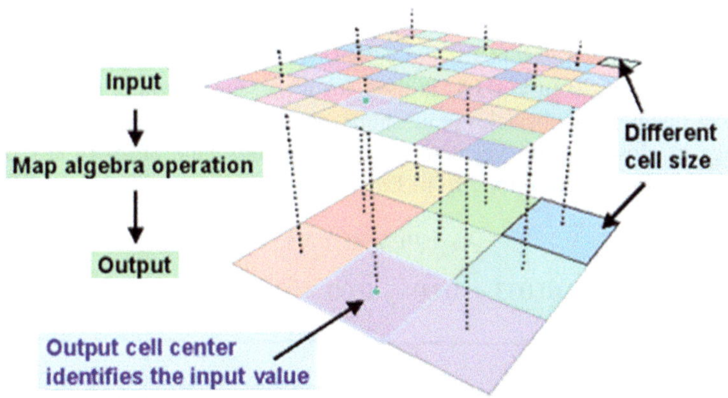

Fig. 4 The workflow of the developed terrain using sub-grid concept

tiles in certain important area and a coarser grid set for quick representation of large areas. A grid merging algorithm was proposed to produce seamless landscape scenes consisting of multiple tiles of different grid layers. First, the original DTM data need to be processed to generate DTM that represented the whole study area. Then the finest grid will be generated at the location where the curbs have been identified through the vectorization process using the ArcGIS 10 syntax below.

$$Con(IsNull(Finer\_grid), [Coarser\_grid], [Finer\_grid])$$

The workflow of the developed terrain using sub-grid concept is shown in Fig. 4. The idea was to set threshold resolutions for different levels. When the level of detail is coarser, the threshold should be larger while when the detailed is needed such as to generated curbs the threshold should be smaller. When visualization starts, the coarser grid set will be used initially to prevent from processing unnecessarily detail terrain model.

2.4 Detection of Closed-to-Earth Vegetation

A general problem in modelling surface runoff in urban areas is the lack of information about the characteristics and distribution of urban soils. Another major disadvantage is the lack of knowledge about the spatial distribution of soil sealing (e.g. grass, bush), which highly influences the amount of surface runoff. In urban hydrology, modellers have to cope with the situation where it is seemingly contradictory but nonetheless possibly can be true especially for those areas where detailed spatial information would be necessary to make useful predictions for flooding (i.e. urban areas).

In this study, the close-to-earth vegetation extraction algorithm encapsulates the points with elevation less than 30 cm excluding the points that has been detected as curbs. Usually the area covered by this kind of vegetation will generate an area (polygon). Once detected, the corresponding points will remain in the DTM but the points will be incorporated with the appropriate Manning's value.

In this study, the Manning value adopted for MIKE21 model is 30 ($n = 0.033$) for the entire surface except for the cells that represent passage buildings. The Manning value of 20 ($n = 0.05$) is used for a passage building, to emulate disturbances due to local obstacles (e.g., shops, bike racks, etc.). For the area with the coverage of close-to-earth vegetations, the Manning value of 29 ($n = 0.034$) is used. This value is determined from Eq. 1.

$$n = (n_b + n_1 + n_2 + n_3 + n_4)m \tag{1}$$

$$n = (0.033 + 0.00 + 0.00 + 0.001 + 0.00)1$$

$$n = 0.034$$

Figure 5 shows the example of roughness coefficient, Manning's value distribution in study area.

3 Case Study

The study area concerns part of the Klang River basin. It is located on the west coast of Peninsular Malaysia, in the Federal Territory of Kuala Lumpur. The Klang River basin is the most densely populated area in the country, with an estimated population of over 3.6 million, which is growing at almost 5 % per year. Although major flood relief works within Kuala Lumpur has been implemented, flooding in the city is still frequent and severe. This causes disruption to various activities in the city, as well as extensive losses suffered by the urban population. The Government of Malaysia, through the Department of Irrigation and Drainage (DID) has commissioned several flood mitigation and river environment enhancement programmes.

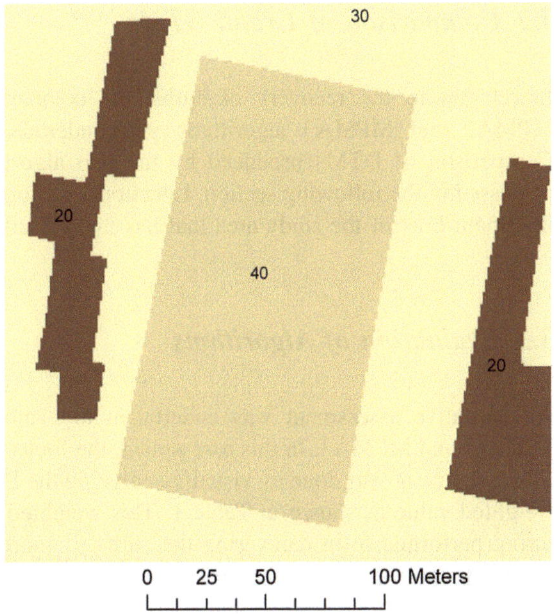

Fig. 5 Example of the distribution of roughness coefficient (Manning's value) in study area

3.1 Model Set-up

The model of the study area contains the river network and urban floodplains for which the 1D/2D commercial software packages MIKE 11/MIKE 21 (i.e., MIKEFLOOD) were utilised. The MIKE 11 model of the river was developed and calibrated previously by DHI Water and Environment in 2004. The drainage channel network and the river are modelled with 1D model (MIKE 11). The calculated subcatchment discharges are introduced as lateral or concentrated inflows into the branches of the 1D model network. The hydrographs generated for each subcatchment are calculated using the NAM and nonlinear reservoir method. Note that the NAM model was used only in the model developed by DHI and has not been used in the present model. Also no sewer model was considered in this current research. The floodplain flows are modelled with the 2D model (MIKE 21). The DTM along the channel network, providing the interface between the coupled 1D-2D models, is set to the bank-full level of the 1D model. Upstream and downstream boundary conditions of the river model are derived from the results of the previous Klang River model and introduced as inflows (upstream) and water level (downstream) for the version of the model used in the present work. The flood extent resulting from the developed models is due to the effects of river-related overbank discharges. For this study, the model set-up is done for flood event on 10th June 2003.

3.2 Comparison of Urban DTMs

In relation to the recovery of curbs, the comparison of DTMs produced by MPMA2 and MPMA3 algorithms was undertaken at two typical locations. Comparison of DTMs produced by the two algorithms is given in Fig. 7 and discussed in the following section. Location 1 to Location 3 in Fig. 6 is situated in the South-East of the study area that has many road networks.

3.3 Evaluation of Algorithms

A qualitative assessment was undertaken to evaluate the DTM developed by MPMA2 and MPMA3. In this assessment, the focus is on the recovering of curbs. The assessment was done by visually analyzing the DTM and giving a mark with a weighted value as shown in Table 1. This weighted value is based on the recapturing performance in recovering the curbs. If more than 75 % of the curbs are captured/recovered the filter is given 1 mark; if 50–75 % of the curbs are captured/recovered the filter is given 2 marks; if 25–50 % of the curbs are captured the filter is given 3 marks and if less than 25 % of the curbs are captured/recovered the filter is given 4 marks. The least total mark gives an indication which filter performs best with respect to the selected criteria.

The DTMs were compared in terms of their efficiency in removing recapturing and recovering the curbs in two locations (see Fig. 6). From the overall analysis of results given in Table 1 it can be noted that the MPMA3 produced better results than MPMA2. This can be also followed by observing Fig. 7, which illustrates the quality of DTMs obtained by two algorithms at the two locations. Further to the improvements described above, the DTM generated by MPMA3 algorithm proved to have better results than its predecessor MPMA2. The DTM generated by MPMA3 appeared to recover almost all curbs that have been left out due to insufficient grid resolution in MPMA2 (1 m) at those locations. MPMA3 managed

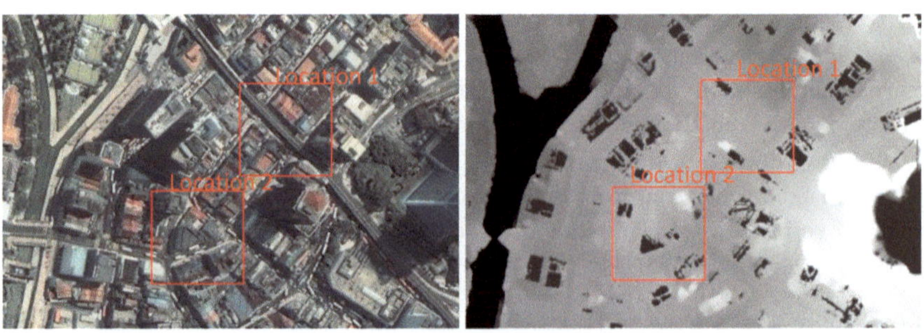

Fig. 6 Sample location within the study area

Table 1 Comparison of performance in curbs capturing and recovering by two algorithms

Indicators	MPMA2	MPMA3
Curbs captured/recovered	3	1
Discontinuity problem	3	1
Total weighted value	6	2

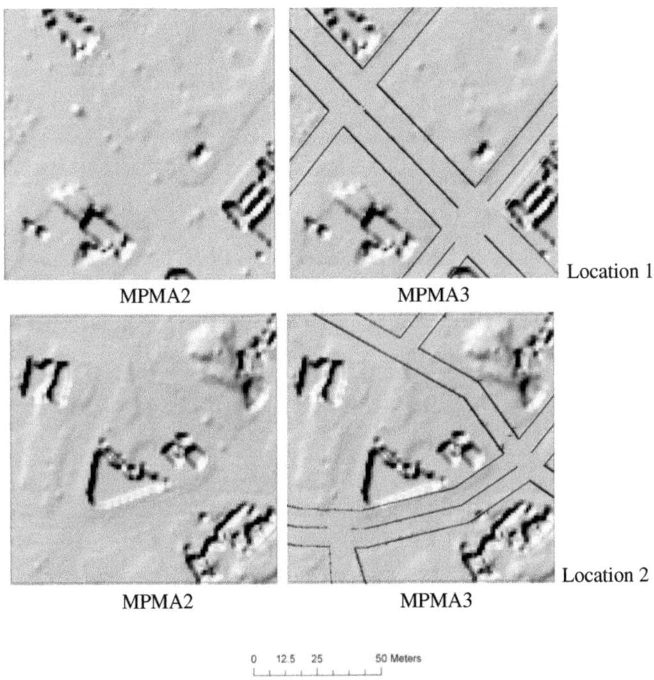

Fig. 7 Comparison of DTM details produced by two algorithms at three locations

to incorporate curbs in all three location using the usage of sub-grid method which in this case was upgraded to 0.2 m (only in location where curbs is been recovered). MPMA2 could capture less than 20 % of the curbs enclosed with discontinuity problem which can be clearly seen in all locations.

3.4 Distribution of Roughness Coefficient (Manning's Value)

Traditionally hydrodynamic models get input of Manning's value manually using estimation based on visual interpretation of available land cover information for model domain. In this study, input of Manning's value is based on the extraction of points during the filtering process. The distribution of roughness coefficient,

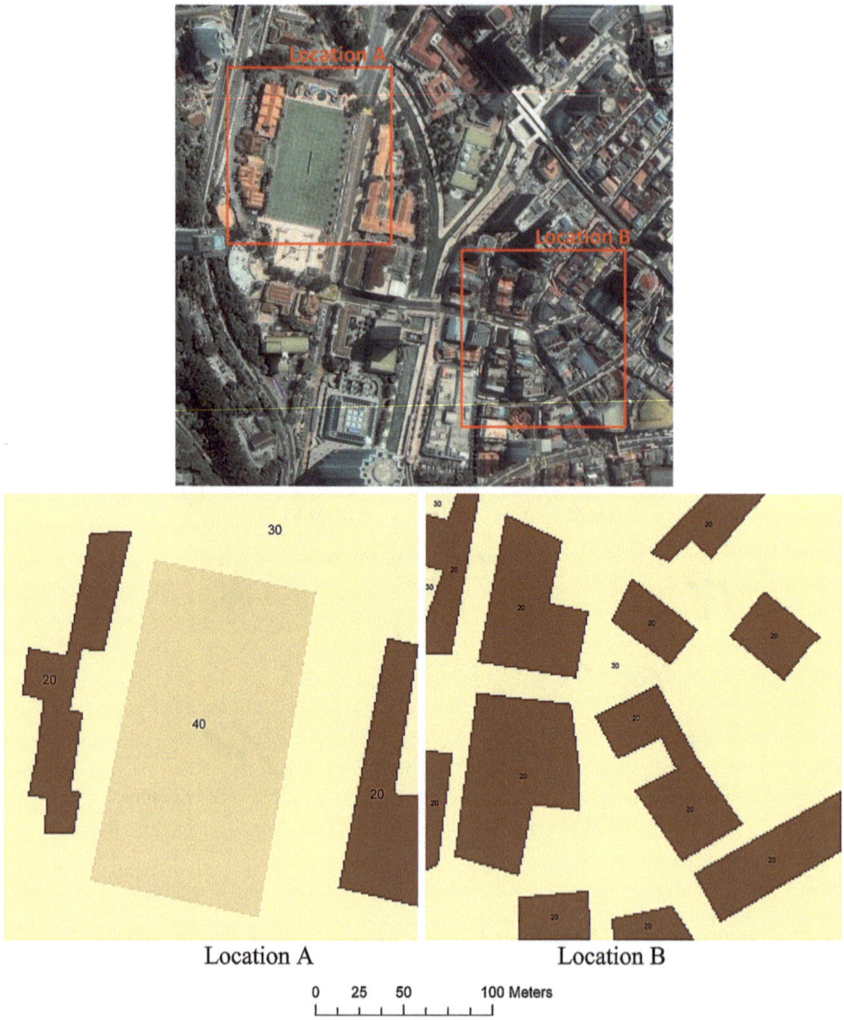

Fig. 8 Distribution of Manning's value within the study area

Manning's value used in the flood models is dominant by three (3) main values which are 20 for the area covered with passage building, 29 for the area covered with close-to-earth vegetation and 30 for anywhere else. In relation to this, the comparison of Manning's value used in the models developed from MPMA2 and MPMA3 algorithms was undertaken at two typical locations (Fig. 8). From Fig. 8, it is shows that value 30 is dominating the study area followed by 20 and 29. It can be obviously understand that in an urban environment there is more impervious area (represent by value 30 and 20) than pervious area (represent by value 29). In Location A, 57 % of the area (0.053 km^2) is represent by value 30, 30 % is represent by value 20 and 13 % is represent by value 29. In Location B, there is only two (2) Manning's value observed which are value 30 (62 %) and 20 (38 %). Close-to-earth vegetation is not present in this location.

3.5 Modelling Framework

The model of the study area contains the river network and urban floodplains for which the 1D/2D commercial software packages MIKE 11/MIKE 21 (i.e., MI-KEFLOOD) were utilised. The MIKE 11 model of the river used in the present work has been developed and calibrated in a previous study carried out by DHI Water and Environment in 2004. The main purpose of that study was to investigate different flood mitigation scenarios. The study also involved a flow/rainfall measuring campaign during the period 1st March 1999—1st April 2003. The time series data used for calibration purposes included nine rainfall events and the results obtained were assessed to be adequate. The details of this work are described in the study report (DHI 2004). The floodplain flows were modelled with 2D (MIKE 21) models which were built using different DTMs generated by MPMA2 and MPMA3 algorithms with spatial resolution of 1 m and coupled with 1D MIKE 11 model. Both 2D models have the same distribution of Manning's roughness coefficients across the modelled domain. In both of the 2D models the buildings (buildings with basement, passage buildings and solid buildings) and elevated road were incorporated in equal way following the previous work (see Abdullah et al. 2011a, b). The 2D model built from MPMA3 algorithm has managed to recover the curbs using the sub-grid method, whereas the MPMA1 algorithm has not been able to do so. Furthermore, the 2D model built from MPMA3 algorithm also replaced the Manning's roughness coefficient for the area containing closed-to-earth vegetation e.g. grass and shrub (height less than 0.3 m) cells with the adjustment value for vegetation. In terms of the coupling with 1D model, the 2D models were set to the bank-full level of the 1D model. Upstream and downstream boundary conditions of the river model were derived from the results of the previously modelled Klang River model and introduced as inflows (upstream) and water level (downstream) into the model used in the present work. The rainfall event that occurred on 10th June 2003 was used in model simulations of the present work. The rainfall data of that event was gathered from two rainfall stations: JPS Wilayah and Leboh Pasar. The recorded rainfall at Station JPS Wilayah was 86 mm during the period of 4 h while the Leboh Pasar station recorded 125 mm for the same duration. To account for spatial variability of rainfall the subcatchments of the 1D model were divided into two groups according to their location and proximity to the two rain gauge stations and as such they were assigned the respective rainfall time series.

3.6 Quality Analysis

In order to determine the potential influence of the filtering algorithms on the resulting DTM, Type I (bare earth misclassified as objects) and Type II (objects misclassified as bare earth) errors in the LiDAR information extraction is calculated for the original Progressive Morphological Algorithm (PMA) and MPMA3. The calculation error is done by calculating the percentage of Type I and Type II errors based on the predominant features in the data set. Five test sites in urban environment within the case study area (see Fig. 9) were chosen because they contained a variety of characteristics that were expected to be difficult for automatic filtering. The datasets included landscapes with densely packed buildings, elevated roads and underpasses, river and bridges, ramp and curbs and large area of bushes and grass. The urban sites were recorded with a point spacing of 1 m.

All points in the datasets were labeled bare earth or object. The definition of what should be considered bare earth is, however, a subjective one. For the purpose of this test, bare earth was defined as the topsoil or any thin layering (asphalt, pavement, etc.) covering it. According to this definition, buildings, bridges, gangways, etc., were treated as objects. Additionally, the bare earth was treated as being a piecewise continuous surface. Therefore, courtyards were also accepted as being part of the bare earth if they were near the surface interpolated between the points on surrounding streets. From the two data sets (PMA and MPMA) five

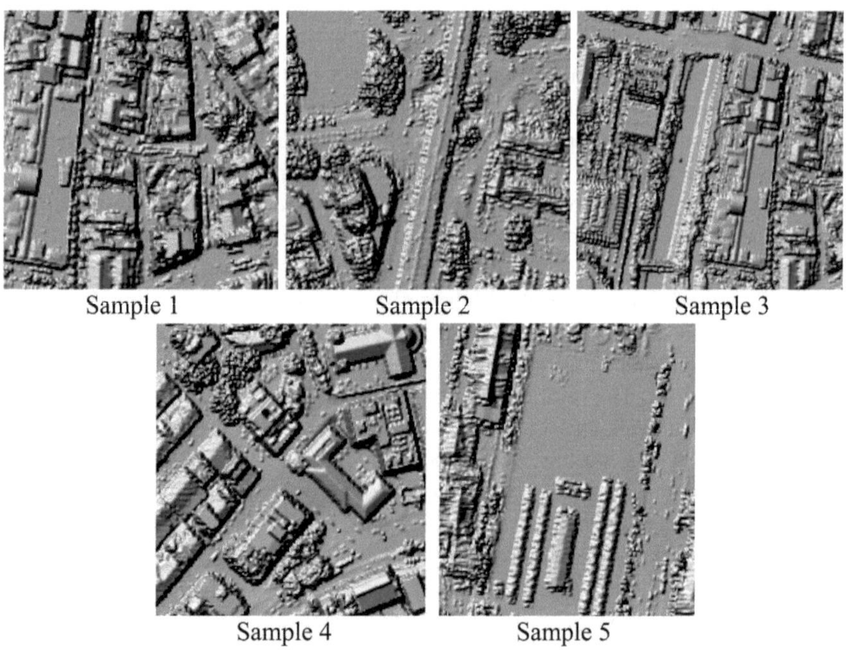

Fig. 9 Test sites

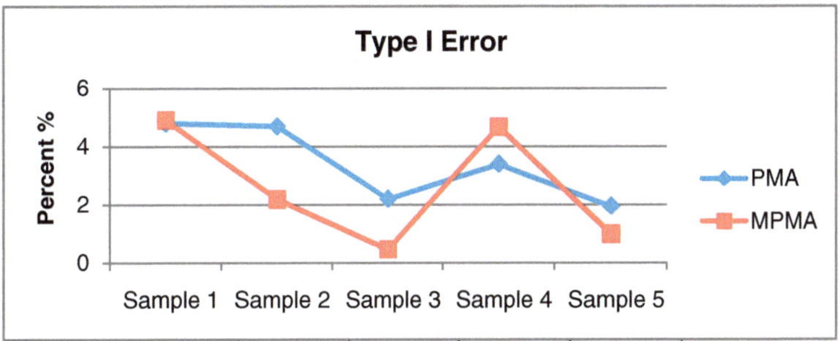

PMA	4.78	4.68	2.19	3.38	1.94
MPMA3	4.89	2.18	0.46	4.65	0.99

Fig. 10 Type I errors over the 5 samples

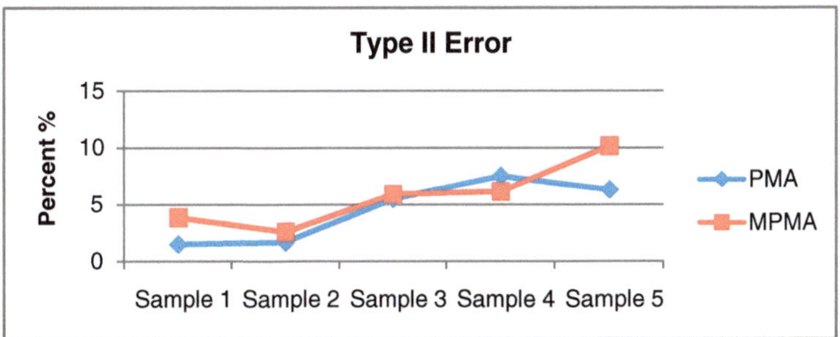

PMA	1.48	1.64	5.44	7.45	6.23
MPMA3	3.82	2.55	5.87	6.07	10.11

Fig. 11 Type II errors over the 5 samples

samples were abstracted. These five samples were representative of different selected important elements in urban environments.

The type I errors calculated are shown in Fig. 10. The type I error presents the number of misclassified bare earth points in a sample as a percentage of all bare earth points in the sample. The developed algorithm does not exhibit large error variations in type I errors (0–5 %) between the sample sites. This is encouraging because it indicates that the developed algorithm is more robust to different landscape types and hence is more reliable.

The type II errors calculated are shown in Fig. 11. The type II error presents the number of misclassified object points in a sample as a percentage of all object points in the sample. The type II errors obtained were relatively small, except for a few sites where the solid buildings and curbs are preserved and rebuilt in the developed algorithm which led to higher errors. Usually, there are more bare earth points then object points which make the impact of type II errors on the total error is small.

3.7 Discussion of Model Results

The model results were analysed in terms of flood depths and extent of flooded area and compared against observations taken at four locations along the streets, shown in Fig. 12. Table 2 indicates the difference between the measured and modelled data at these locations.

The differences between the measurements and results from the 2D model built from MPMA3 were found to be between 0.4 and 4.4 %. In terms of the differences between the measurements and results from the 2D model built from MPMA2 algorithm, this was found to be in the range between 0.2 and 6.7 %. In Jalan Raja (1) and Tun Perak (3), the differences between the measurements and results from the 2D model built from MPMA3 seen to be decreased and become closer to the measurement. In Leboh Ampang (2), the results shows some increasing of flood depth by 0.011 m by the results from the 2D model built from MPMA3 while in Jalan HS Lee (4) the result remain the same for both model. The increasing result shows in Leboh Ampang (2) can be attributed to the effects of the curbs recovery along the streets which encourage the flood water to be gathered more at the center of those streets. The effect of close-to-earth vegetation in this area is found to be less significant as the covered area is relatively small compared to urban surface (streets, pavements).

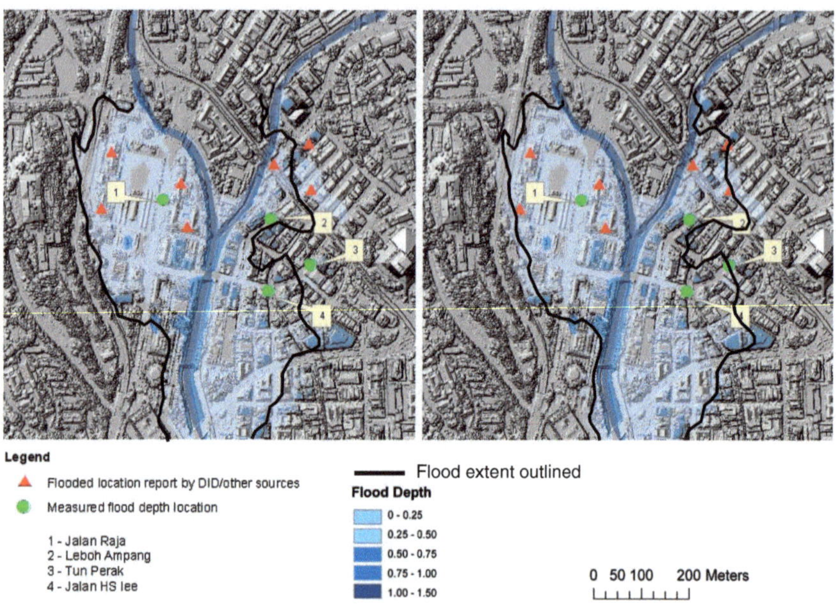

Fig. 12 Modelled and observed flood locations for the 10th June 2003 rainfall event. Images illustrate predictions by models with DTMs generated from MPMA1 (*left*) and MPMA2 (*right*) algorithms. Observed locations are represented with *triangles* and *circles*

Table 2 The difference between the measured and modelled data at four locations along the streets

Location Flood depth (m)	Jalan Raja (1)	% Error	Leboh Ampang (2)	% Error	Tun Perak (3)	% Error	Jalan HS Lee (4)	% Error
Measured	1		1.2		1		1	
1D/2D model with MPMA3 DTM	0.956	4.4	1.209	0.7	0.996	0.4	0.995	0.5
1D/2D model with MPMA2 DTM	0.933	6.7	1.198	0.2	0.984	1.6	0.995	0.5

Model results using the DTM from MPMA2 algorithm; Model results using the DTM from MPMA3 algorithm.

For the overall study area (1 km^2), the results from the 2D model built from MPMA2 algorithm show that 30 % of that area (or 300,000 m2) is flooded. In terms of the results obtained from the 2D model built from MPMA2 algorithm, the flooded areas (315,000 m^2) is observed to be increased by 5 % from the area in MPMA2.

From the analysis of computed flood velocities, it can be observed that the results from the 1D/2D model that uses the DTM from the MPMA3 algorithm shows the flood water has been controlled by the curbs so that it will flow along the street. This results which correspond better to reality as the DTM better represents the streets characteristics (Fig. 12). In the 1D/2D model that use the DTM from the MPMA2 algorithm the water is seen to flow without certain guidance through the street as the curbs cannot be captured by 1 m grid resolution. This is also evident from observations of velocity vectors given in Fig. 13 (refer also Fig. 7). As in Location A, (refer Fig. 8), it is seen that in MPMA3 model, where the surface is covered by the close-to-earth vegetation, which in this case, grass, the flood is observed to flow slower than the same area in MPMA2 model. The difference in results indicate that micro objects especially the ones that correspond to the street characteristic such as curbs need to be carefully processed, otherwise it can generate misleading results.

Traditionally hydrodynamic models are provided with Manning's values manually using estimates based on a visual interpretation of available land cover information for the model domain. In this study, the input of Manning's value is based on the extraction of points during the filtering process. The distribution of the Manning roughness number used in the flood models is dominated by three main values which are 20 for the area covered with passage building, 29 for the area covered with close-to-earth vegetation and 30 for anywhere else. In relation to this, the comparison of the Manning's number used in the models developed from the MPMA2 and MPMA3 algorithms was undertaken at two typical locations. Figure 8 shows that the value 30 is dominant for the study area followed by 20 and 29. It is obvious that in an urban environment there is more impervious area (represent by the values 30 and 20) than pervious area (represent by the value 29). In Location A, 57 % of the area (0.053 km^2) is represented by the value 30, 30 %

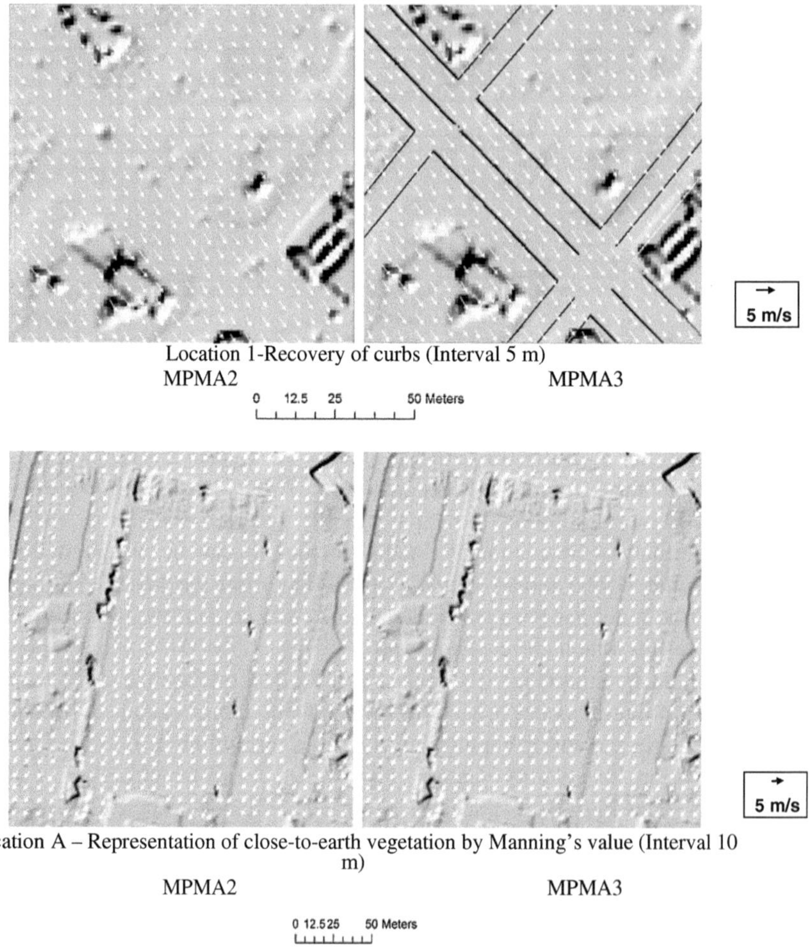

Fig. 13 Comparison of velocity vectors and DTMs generated from MPMA2 (*left* image) and MPMA3 (*right* image) at two sample location (refer Figs. 6 and 7)

is represented by the value 20 and 13 % is represented by the value 29. In Location B, there are only two (2) Manning's values observed, 62 % of values 30 and 38 % of value 20. Close-to-earth vegetation is not present at this location.

From all the evaluations above, it is seen that the developed algorithms show better result in term of flood depth, flood extent and flood velocity compared with the existing filtering algorithm. The combination of the filtering process, the incorporated of solid buildings and curbs in the DTM and the representative of areas with appropriate roughness coefficient (Manning's value) in the developed algorithm really contributed to a better urban flood models which represent as close as the realistic urban surface.

4 Conclusion

From the work described here, further modification has been made to the filter algorithms, in order to recover the curbs and on the DTM. In addition, the close-to-earth vegetation has been removed and represented with the appropriate Manning's coefficient value. Model simulations were performed using the data from the flood event that occurred on 10th June 2003 in order to make an evaluation of the new MPMA3 algorithm. The model results were analysed and compared with those using MPMA2 in terms of flood depths and flood extent. From the analysis of results, it can be observed that some improvement has been achieved in the context of flood depth while in flood extend, the results remain almost the same for both models. The work described in this chapter demonstrates that the weakness in capturing micro features such as curbs due to insufficient grid resolution and discontinuities present in urban environments especially in streets network can play a significant role in diverting and directing shallow water flows, therefore the selection of an appropriate LiDAR filtering algorithm and the grid resolutions plays an important role in flood modelling work.

References

Abdullah AF, Vojinovic Z, Price RK, Aziz NAA (2011a) A methodology for processing raw LiDAR data to support urban flood modelling framework. J Hydroinformatics 14(1):75

Abdullah AF, Vojinovic Z, Price RK, Aziz NAA (2011b) Improved methodology for processing raw LiDAR data to support urban flood modelling—accounting for elevated roads and bridges. J Hydroinformatics 14(2):253

Chen AS, Hsu MH, Teng WH, Huang CJ, Yeh SH, Lien WY (2005) Establishing the database of inundation potential in Taiwan. Nat Hazards 37(1–2):107–132

Chen Q (2007) Airborne LiDAR data processing and information extraction. Photogram Eng Remote Sens 73:109–112

DHI Water and Environment (M) Sdn. Bhd. (2004) Klang river basin environment improvement and flood mitigation project (storm water management and road tunnel—smart), Final report to Government of Malaysia

Djordjevic S, Prodanovic D, Maksimovic C, Ivetic M, Savic DA (2005) SIPSON—simulation of interaction between pipe flow and surface overland flow in networks. Water Sci Technol 52(5):275–283

Fortune S (1985) A sweepline algorithm for voronoi diagrams. Algorithmica 2(2):153–174

Flood M (2001) Laser altimetry—from science to commercial lidar mapping. Photogram Eng Remote Sens 67(1209–11):1213–1217

Garcia-Navarro P, Brufau P (2006) Numerical methods for the shallow water equations: 2D approach. In: Knight DW, Shamseldin AY (eds) River basin modeling for flood risk mitigation. Taylor & Francis, London, pp 409–428

Haile AT, Rientjes TH (2005) Effects of LiDAR DTM resolution in flood modelling: a model sensitivity study for the city of Tegucigalpa, Honduras. ISPRS workshop 'laser scanning 2005', Enschede

Hunter NM, Bates PD, Horritt MS, Wilson MD (2007) Simplified spatially-distributed models for predicting flood inundation: a review. Geomorphology 90:208–225

Hunter NM, Bates PD, Neelz S, Pender G, Villanueva I, Wright NG, Liang D, Falconer RA, Lin B, Walle S, Crossley AJ, Mason DC (2008) Benchmarking 2D hydraulicmodels for urban flooding. Wat Manage 161(1):13–30

Kuiry SN, Sen D, Bates PD (2010) A coupled 1D-quasi 2D flood inundation model with unstructured grids. J Hydraul Eng 136(8):493–506

Lemmens M (2007) Airborne lidar sensors. GIM Int 21(2):0

Mark O, Weesakul S, Apirumanekul C, Aroonnet SB, Djordjevic S (2004) Potential and limitations of 1D modelling urban flooding. J Hydrol 299(3–4):284–299

Norman E (2007) Surface-sewer coupling and detailed elevation models for accurate urban drainage modelling. In: Proceedings Cost Session Aquaterra Conference 2007, Special aspects of urban flood managements, pp 183–196

Price RK, Vojinovic Z (2011) Urban hydroinformatics: data, model and decision support for integrated urban water management. IWA, London

Romano ME (2004) Innovation in LiDAR processing technology. Photogram Eng Remote Sens 70:1202–1206

Sangster C (2002) Validating LiDAR—evaluating LiDAR accuracy using GPS. Applied Geomatics Research Group, Centre of Geographic Science, Lawrence town, Nova Scotia

Schmitt TG, Thomas M, Ettrich N (2004) Analysis and modelling of flooding in urban drainage systems. J Hydrol 299:300–311

Schmitt TG, Schilling W, Sægrov S, Nieschulz KP (2002) Flood risk management for urban drainage systems by simulation and optimisation. In: 9th ICUD conference, Portland

Vojinovic Z, Tutulic D (2009) On the use of 1D and coupled 1D–2D modelling approaches for assessment of flood damages in urban areas. Urban Wat J 6(3):183–199

Vojinovic Z, Seyoum SD, Mwalwaka JM, Price R (2011) Effects of model schematisation, geometry and parameter values on urban flood modelling. Water Sci, Technol 63

Potential of Multiscale Texture Polarization Ratio of C-band SAR for Forest Biomass Estimation

Latifur Rahman Sarker, Janet Nichol and Ahmad Mubin

Abstract Estimation of forest biomass is still a challenging task over large areas because of the saturation problem of remote sensing data as well as the environmental, topographic, and biophysical complexity of forest ecosystems. However, Synthetic Aperture Radar (SAR) is still one of the most attractive choices for the estimation of biomass or carbon storage capacity of vegetation due to its sensitivity to plant canopy structure, and new SAR sensors have attracted greater interest as they are able to provide data with varying spatial resolutions, polarizations, and incidence angles. This research investigates the potential of C-band dual polarization (HH and HV) SAR (Radarsat-2) imagery for forest biomass estimation using different combinations of raw backscattering (intensity), polarization texture parameters and texture polarization ratios. Several models have been developed between field biomass and SAR signatures using stepwise multiple regression. Results indicate that SAR intensity images (C-HV and C-HH) and the ratio of intensity data (C-HV/C-HH) have relatively low potential ($r^2 = 0.20$) for biomass estimation. However, the SAR polarization (C-HV and C-HH) texture parameters were found to be effective and about 82 % ($r^2 = 0.82$ and RMSE = 28.68 t/ha) of the variability in the field data (forest biomass up to 360 t/ha) was explained by the model. Further improvement of the estimation was achieved ($r^2 = 0.90$ and RMSE = 21.55 t/ha) using the texture polarization ratio (C-HV/C-HH). The outcomes suggest that a clear improvement in forest biomass estimation can be obtained using the texture parameters of dual polarization

L. R. Sarker (✉) · A. Mubin
Department of Geoinformation, Faculty of Geoinformation and Real Estate, Universiti Teknologi Malaysia, Johor Bahru, Malaysia
e-mail: lrsarker@yahoo.com

L. R. Sarker
Department of Geography and Environmental Studies, University of Rajshahi, Rajshahi, Bangladesh

J. Nichol
Department of Land Surveying and Geo-Informatics, The Hong Kong Polytechnic University, Hung Hom, Hong Kong

A. Abdul Rahman et al. (eds.), *Developments in Multidimensional Spatial Data Models*, Lecture Notes in Geoinformation and Cartography,
DOI: 10.1007/978-3-642-36379-5_5, © Springer-Verlag Berlin Heidelberg 2013

C-band SAR and more improvement can be achieved using the ratio of texture parameters, as this combines the advantages of both texture and ratio.

Keywords RADARSAT-2 · Texture measurement · Forest biomass · Carbon storage capacity and saturation level

1 Introduction

Forest biomass estimation is still a challenging task and this is especially true in tropical and sub-tropical forest because of its complicated stand structure and species heterogeneity (Foody et al. 2003; Lu 2006). However, accurate estimation of regional and global forest biomass is essential for understanding and monitoring ecosystem response to climate change (Brown et al. 1989; Ranson and Sun 1994) for greenhouse gas inventories, terrestrial carbon accounting and climate change modeling studies (Dobson et al. 1992; Muukkonen and Heiskanen 2007). Satellite sensors now offer an effective method for forest biomass and carbon inventory at local, regional and global scales (Le Toan et al. 1992; Rosenqvist et al. 2003; Foody et al. 2003). Many studies have been carried out for the estimation of forest biomass using data from different types of satellite sensors such as optical (Thenkabail et al. 2004; Zheng et al. 2004; Muukkonen and Heiskanen 2007; Sarker and Nichol 2011), SAR (Le Toan et al. 1992; Dobson et al. 1995; Kuplich et al. 2005; Champion et al. 2008; Sarker et al. 2012), Light Detection and Ranging (LIDAR) (Nelson et al. 1988; Bortolot and Wynne 2005) and multisensory (Hyde et al. 2007; Sarker and Nichol 2011), but no universal and transferable technique has been developed so far to quantify forest biomass/forest carbon because of the saturation problem of remote sensing data as well as the environmental, topographic, and biophysical complexity of forest ecosystems (Foody et al. 2003). However, for forest biomass estimation, the most promising type of sensor appears to be Synthetic Aperture Radar (SAR) due to its sensitivity to forest structure (Harrell et al. 1997) as well as all-weather capability (Dobson et al. 1992; Le Toan et al. 1992).

Almost all past studies have indicated that L-band cross polarization (L-HV) is the most suitable available choice (as no P-band space borne SAR is planned) and the C-band SAR is a less suitable choice because of the problem of saturation level of SAR data (Le Toan et al.1992; Harrell et al. 1997). However, the previously assumed saturation levels are based on mono-frequency and mono-polarization SAR data (Dobson et al. 1995; Foody et al. 1997) and saturation levels are not constant. The saturation level of SAR backscatter varies as a function of radar wavelength, polarization and incidence angle (Le Toan et al.1992; Imhoff 1995) as well as many physical factors of the study area. The new generation space borne SAR sensors i.e. L-band PALSAR (no longer operational) C-band RADARSAT-2, X-band Terra SAR and Cosmo Skymed are more attractive than previous generation space borne SAR sensors i.e. JERS-1, ERS-1/2 and RADARSAT-1, and shows great potential for biomass estimation and carbon storage capacity of vegetation as these sensors are able to provide data with varying spatial resolutions, polarizations, and incidence angles. However, the improvement of forest biomass estimation depends not only on the SAR data but also requires efficient SAR data processing (Imhoff 1995), and one of the most important spatial data processing techniques is texture processing (Luckman et al. 1997; Champion et al. 2008; Sarker et al. 2012).

Texture is an important characteristic used to identify objects or regions of interest in any image (Haralick et al. 1973) and studies have shown that in most cases, texture, not intensity, is the most important source of information in high-resolution radar images (e.g., Dobson et al. 1995; Dell'Acqua and Gamba 2003), since it can maximize the discrimination of spatial information independently of tone, thereby potentially increasing the range of biomass to higher levels (Luckman et al. 1997). Promising result have been demonstrated for the use of texture measurement for land cover and vegetation classification (Milne and Dong 2002) as well as biomass estimation using both optical (Sarker and Nichol 2011) and SAR data (Luckman et al. 1997; Champion et al. 2008; Sarker et al. 2012). But SAR image texture and its relation with forest biomass has not yet been fully investigated (Luckman et al. 1997; Kuplich et al. 2005), probably due to the difficulty of selecting appropriate texture measures along with window size (Chen et al. 2004) as well as the previous unavailability of high resolution satellite borne SAR data.

However, considering the necessity of new techniques for the estimation of forest biomass as well as the lack of investigation of texture measurement in combination with the new advanced SAR sensor (RADARSAT-2), the objective of this research is to explore the potential of advanced SAR sensor (RADARSAT-2) for the estimation of forest biomass using different types of image combinations and image processing technique (such as texture measurement). The RADARSAT-2 SAR was selected because it provides dual polarization, suitable incidence angle, and high spatial resolution, as well as based on the observation and findings of previous studies (Luckman et al.1997; Castro et al. 2003) that C-band SAR can provide excellent texture properties for biomass estimation when it is acquired at a high spatial resolution.

2 Study Area and Data

2.1 Study Area Selection

The study area for this research is the Hong Kong Special Administrative Region (Fig. 1). The total land area of Hong Kong is 1,100 km^2, and approximately 40 % of the total area is designated as Country Parks which are reserved for natural vegetation under the management of the Agriculture, Fisheries and Conservation

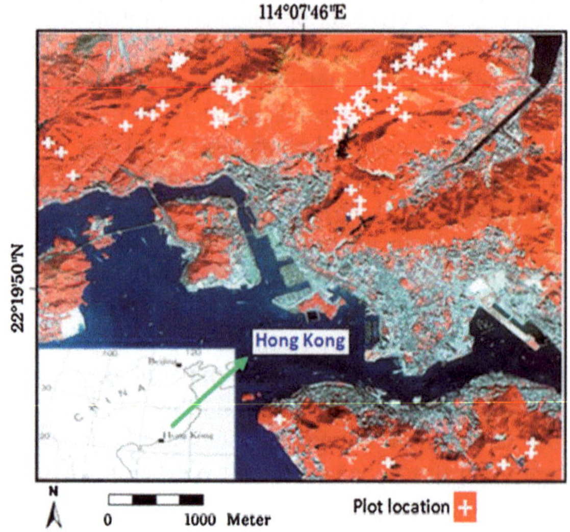

Fig. 1 Study area and distribution of field plots

Department (AFCD). The native sub-tropical evergreen broad leaf forest of Hong Kong has been replaced by a complex patchwork of regenerating secondary forest in various stages of development, and plantations.

2.2 Data

A dual polarization (HV and HH) C-band fine-beam image from the RADARSAT-2 SAR satellite with approximate resolution of 10.4–6.8 m in range and 7.7 m in azimuth directions was used in this study. Data was collected on the 16th of January 2009 with an incidence angle of 34.2°.

3 Methodology

The overall methodology of this study is presented in Fig. 2, however, a short description of the different steps of the methodology is provided in the following.

3.1 Allometric Model and Field Biomass Estimation

Due to the lack of an allometric model for converting the measurable tree parameters to actual biomass, the harvesting of a large sample was done by selecting the dominant tree species comprising a total of 75 trees in 4 DBH classes

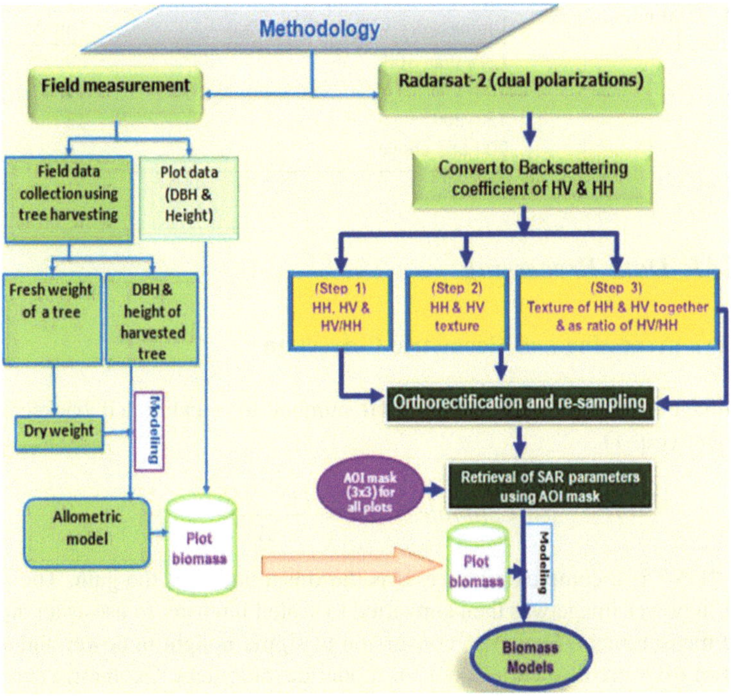

Fig. 2 Overall methodology

(less than 10 cm, 10–15 cm, 15–20 cm, 20 cm and above) and standard procedures were followed for tree harvesting (Overman et al. 1994; Ketterings et al. 2001). Twelve regression models used by previous researchers (Brown et al. 1989; Overman et al. 1994; Arevalo et al. 2006) were tested in order to find the best fit using dry weight (DW) as the dependent variable and diameter breast height (DBH) and height as independent variables in different combinations. The final model $\left(\widehat{DW} = \mathrm{e}^{-2.057}\mathrm{DBH}^{2.289}\mathrm{e}^{\mathrm{mse}/2}\right)$ was selected considering all parameters including correlation coefficient (r), coefficient of determination (r^2), adjusted coefficient of determination (adjusted r^2), and RMSE. The detailed description of the development of allometric model can be found in Sarker and Nichol (2011).

Forty-five sample plots covering a wide variety of tree stands were selected using purposive sampling. The center of 15 m radius circular plots was determined with a Leica GS5 + Global Positioning System (GPS) and all sample plots were positioned within a homogenous area of forest. Both DBH (1.3 m above ground) and tree height were measured for all trees within the circular plot region but trees with diameter less than 2.5 cm DBH were not included. Finally using the measured parameter DBH, the biomass of each tree and subsequently biomass of all trees in a plot were estimated (Table 1) using the allometric model developed for this study area.

Table 1 Field biomass/dry weight distribution

Biomass/dry weight class (t/ha)	Number of plots
Less than 100	9
100–150	17
150–200	13
200 and above	6

3.2 SAR Data Processing

3.2.1 Pre-processing and Geometric Correction

SAR image was converted from digital number to a calibrated backscattering coefficient (Eq. 1).

$$Calibrated\,value = \frac{DN^2 + B}{A} \tag{1}$$

where "DN" is a complex number, B is the offset and A is the gain. The sigma-nought in power image was then converted to scaled intensity to use as an input for texture measurement. However, conversion to sigma-nought in power and sigma-nought in dB were also carried out for model development. Geometric correction was done using 41 well distributed ground control points, a high resolution (10 m) DEM, a SPOT-5 image (as reference image), The RMS error in X and Y was 0.38 and 0.31 and the overall error was 0.23 pixel. All images were resampled to a 10 m pixel size using nearest neighbor resampling.

3.2.2 Texture Analysis

Several methods and techniques for describing texture, based on statistical models have been developed but there is no guideline as to which methods are suitable for a particular situation, as texture measurement is complex and varied based on many factors. Therefore, in order to find some potential texture parameters which are correlated with field biomass but uncorrelated or less correlated with other independent variables, three categories of texture algorithms (Sarker et al. 2012) were tested both individually and in combination. The first is the grey level co-occurrence matrix (GLCM) (Haralick et al. 1973) along with some grey level difference vector (GLDV) based texture measurements. The second is the sum and difference histogram proposed by Unser (1986) as an alternative to the usual co-occurrence matrices used. The third group is model-based log form texture parameter estimation, reported by Oliver (1993) and Oliver and Quegan (2004). All texture measurements were performed using small to medium window sizes (from 3×3 to 9×9) and four texture measurement directions (0, 45, 90 and 135°) were specified for the processing. The 3 classes of texture measurements used in the research are named TEX for GLCM based texture measurement, HISTEX for Sum and difference histogram based texture parameter, SARTEX for model-based log form texture parameter and ALLTEX for all texture classes together.

3.3 Statistical Analysis

Forest biomass estimation model was developed between field biomass and remotely sensed data using a stepwise multiple-linear regression approach. From each image, the mean SAR response (intensity/SAR texture parameters/SAR texture polarization ratio) of 45 field plots was extracted using an area of interest (AOI) mask of 3 × 3 pixels. The image parameters were used as independent variables and the plot biomass components as the dependent variables. To avoid multicollinearity and overfitting problems as well as to ensure finding the best fit model, all models were tested considering statistical parameters, namely coefficient of determination (r^2), adjusted r^2, RMSE and p-level, Beta coefficient, Std. Error of B, tolerance, variance inflation factor, and condition index.

4 Results and Analysis

The analysis was performed in 3 steps based on the data combinations and processing techniques

4.1 Step-1: Intensity Images

Three models were developed to investigate the relationship between field biomass and simple radar intensity (HH, HV and ratio of HV/HH). Very poor accuracy (r^2) was obtained from the intensity data (HH = 0.05, HV = 0.08 and HV/HH = 0.20) but the ratio of HV/HH intensity data showed better performance ($r^2 = 0.20$) than any single polarization data, although the accuracy (ca. 20 %) was still far below the acceptable level. The poor results obtained from the SAR intensity data was not unexpected and attributed probably because of the three important facts i.e. (1) the speckle noise in the raw (intensity) SAR data, (2) the complex forest structure and rugged topography of the study area, and (3) the most importantly the biomass of the plots is beyond the saturation level of the raw data.

4.2 Step-2: Texture Parameters for HH and HV Polarized Image

Four models were developed between the field biomass and texture parameters of HH polarization SAR data (TEX, HISTEX, SARTEX and ALLTEX). The accuracy (r^2) of HH polarized texture data was very similar to intensity of HH data. Among the 4 models none was found to be effective for the estimation of forest biomass, although TEX ($r^2 = 0.20$) and ALLTEX ($r^2 = 0.20$) showed better performance than HISTEX ($r^2 = 0.14$) and SARTEX ($r^2 = 0.13$). The poor results can be explained by the fact that HH polarized data is known to be less sensitive to forest biomass (Rauste et al. 1994; Kasischke et al. 1995) and adding more parameters using different texture algorithms was still unable to significantly improve the sensitivity of HH polarization data.

Similar to HH polarized data processing, four models were developed between the field biomass and texture parameters of HV polarization SAR data (TEX, HISTEX, SARTEX and ALLTEX). All texture models (TEX, HISTEX, SARTEX and ALLTEX) showed better performance (Fig. 3) and significant improvement ($r^2 = 0.59$) was observed using ALLTEX HV polarized data compared to intensity ($r^2 = 0.20$) and texture of HH polarized data ($r^2 = 0.20$). This achievement can be illustrated by the fact that HV polarized SAR data has better sensitivity for biomass estimation (Le Toan et al. 1992; Dobson et al. 1992) and this sensitivity can be improved using image processing techniques including texture measurement (Kuplich et al. 2005; Sarker et al. 2012).

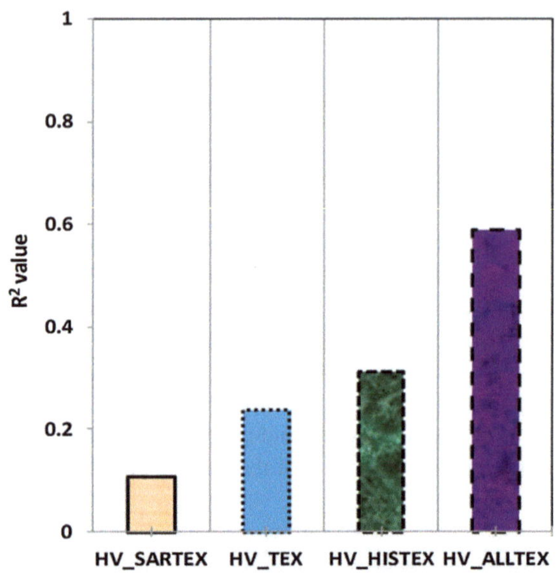

Fig. 3 Accuracy of different models using HV texture parameters

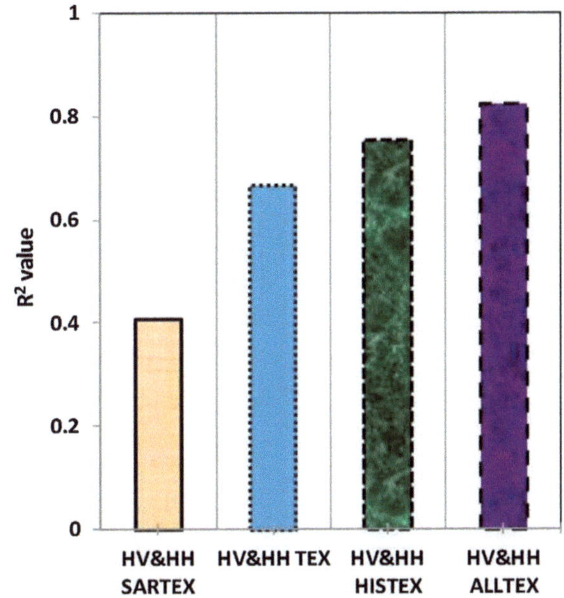

Fig. 4 Accuracy of different models using HH and HV texture parameters

4.3 Step-3: Combination of Texture Parameters for HV and HH Polarized Data

The texture parameters from both polarization (HH and HV) were also used together in the model without ratio (HH and HV) and with ratio (HV/HH) in order to investigate the potential of complementarity information from both polarization (HH and HV) for the forest biomass estimation.

It is clear from the results that the accuracy of the biomass estimation increased (Fig. 4) significantly using both polarization textures (HH and HV) together in the model, though the magnitude of the improvement was different with respect to texture algorithms. Using the HH and HV texture parameters together, the accuracy (r^2) of 0.82 was obtained compared with previous highest accuracy of 0.20 from intensity, 0.20 from texture of HH and 0.59 from texture of HV. The performance of the individual texture modules varied and the lowest accuracy ($r^2 = 0.41$) was achieved using SARTEX (Fig. 4), while highest accuracy ($r^2 = 0.82$) was obtained from ALLTEX (Fig. 4) and the second highest accuracy ($r^2 = 0.75$) from HISTEX (Fig. 4). The best fit model used 7 independent variables (Mean Euclidean distance_3 × 3_HV + Mean Euclidean distance_3 × 3_HH + Mean_3 × 3_HV + Original HV + Inverse differences_9 × 9_HV + Contrast _9 × 9_HH + Kurtosis_5 × 5_HH) from all texture modules together. The model, intercept, and all variables were significant and no multicollinearity effect was suspected (considering the Tol, VIF, and CI).

Fig. 5 Accuracy of different models using ratio of HV and HH texture parameters

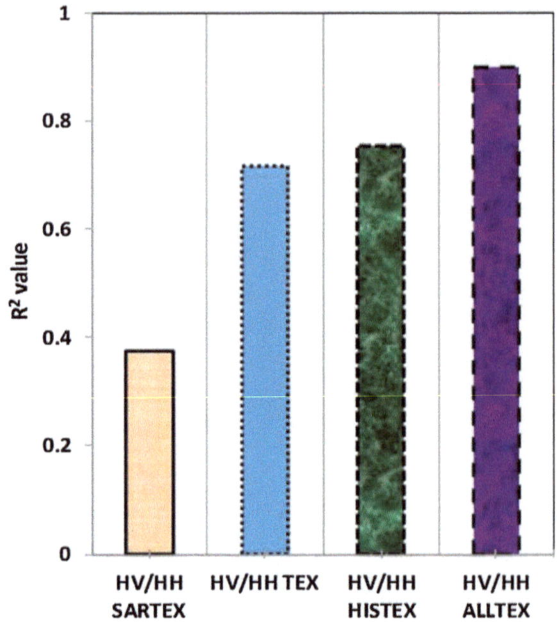

Fig. 6 Comparison of accuracy among intensity (HH, HV, and HV/HH), HH texture, HV texture, combination of HV and HH texture, and ratio of HV and HH texture parameters

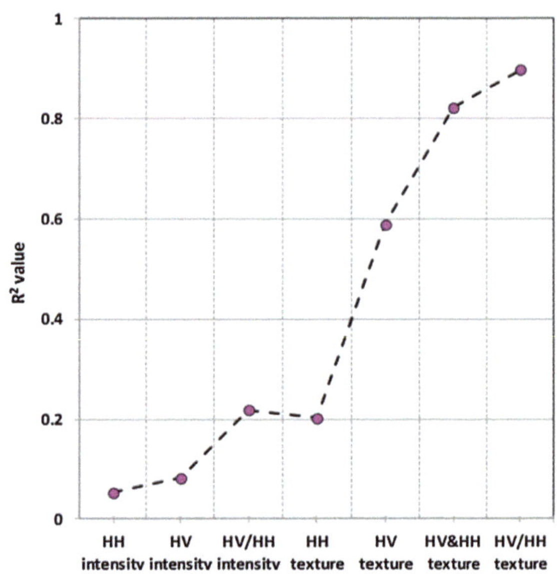

On the other hand, significant improvement of forest biomass estimation was achieved using the ratio of texture images compared to intensity, HH texture, HV texture and combination of HH and HV texture images (Figs. 5, 6).

The accuracy (r^2) of 0.90 was achieved using the texture polarization ratio. The increase in accuracy (Fig. 6) is 350, 350, 52.54, 09.49 %, compared to the best accuracy of intensity, HH texture, HV texture and combine HH and HV texture images respectively. Using the ratio of texture images, the performance of individual textures was variable (Fig. 5 and Table 2), however, highest accuracy was achieved using all texture modules together (ALLTEX), which indicates that the different texture modules consistently provide complimentary information for biomass estimation. Among the four different models (Table 2), the ALLTEX texture model achieved a very significant improvement with an accuracy of 0.90 (r^2) and RMSE 21.55 t/ha. This model used 8 independent variables (Model-1 in Table 2) and the model, intercept and all variables are significant with no apparent multicollinearity effect.

5 Discussion and Conclusion

This research achieved a significant improvement of forest biomass estimation using the C-band SAR, although C-Band has previously been considered the least suitable SAR wavelength for biomass estimation because of the saturation problem. However, this achievement can be explained by the high quality SAR data, newly collected field biomass data, in combination with image processing technique applied. Our first attempt, using the simple intensity of HH, HV and the ratio of HV/HH, obtained poor results probably because of the relatively high biomass level in the study area and inability of intensity data to predict such high levels biomass because of speckle noise. This result is in agreement with the findings of Dobson et al. (1995) who found that retrieval of dry biomass from simple backscattering coefficient or multifrequency ratios was ineffective. The results also comply with Foody et al. (1997) who found no significant relationship between SAR backscatter in the six raw SAR data configurations (CHH, CHV, CVV, LHH, LHV, LVV) and forest biomass.

Despite the use of many texture parameters from HH polarization in the model, the accuracy was very low. This poor performance for texture data of HH polarization is in agreement with other researchers who found that HH is less sensitive for biomass estimation (Rignot et al. 1994; Kasischke et al. 1995). However, our results using texture parameters with HV polarization also agree with previous work (Le Toan et al. 1992). who found that HV polarization of SAR data is very effective for biomass estimation Nevertheless, it is still clear that only slightly above 59 % accuracy can be achieved using the texture with HV polarization. This drawback is in agreement with the findings of Ranson and Sun (1994) who found that the HV channel alone cannot make significant improvement for biomass estimation.

When texture parameters from both image channels (HH and HV) were used together for analysis, the accuracy improved substantially ($r^2 = 0.82$). This improvement agrees with previous studies (Kasischke et al. 1995; Hyde et al.

Table 2 Results obtained from ratio of texture parameters (HV/HH) using different models

Data	Model fitting parameters				Fitting parameters for intercept and variables						
	R^2	$A.R^2$	RMSE	p-level	Variables and intercept	B	Std. Err. of B	p-level	Tol	VIF	CI
Model-1 (ALLTEX)	0.899	0.0877	21.55	0.005	Intercept	73.20	18.31	0.00030	–	–	–
					Variance_3	99.33	9.39	0.00005	0.74	1.35	2.04
					Angular second moment_7	62.54	8.24	0.00005	0.54	1.86	2.72
					Homogeneity _3	0.45	0.07	0.00005	0.27	3.69	3.81
					Inverses differences_3	−20.70	4.18	0.00001	0.25	4.07	4.57
					Kurtosis 7	−13.09	3.73	0.00123	0.93	1.07	6.81
					Standard deviation_3	−7.84	2.20	0.00104	0.28	3.56	9.81
					GLDV Mean_5	45.17	12.99	0.00135	0.16	6.22	10.10
					Contrast_7	−11.90	5.08	0.02472	0.26	3.83	17.35
Model-2 (HISTEX)	0.755	0.724	33.63	0.000	Intercept	103.25	13.90	0.00005	–	–	1.00
					Variance_3	63.55	13.81	0.00004	0.56	1.79	1.85
					HV/HH_ ratio	8.50	2.51	0.00165	0.92	1.09	2.14
					Energy_9	509.64	184.73	0.00879	0.55	1.81	2.43
					HV/HH_ ratio	−1.26	0.52	0.02070	0.97	1.03	3.97
					Kurtosis_7	−11.23	5.45	0.04621	0.98	1.02	6.56
Model-3 (SARTEX)	0.377	0.347	53.71	0.005	Intercept	108.39	12.50	0.00005	–	–	–
					HH/HV_ratio	15.18	3.72	0.00019	0.99	1.01	1.69
					VA_3	7.59	2.33	0.00221	0.99	1.01	2.68
Model-4 (TEX)	0.716	0.654	36.21	0.000	Intercept	93.53	22.86	0.00023	–	–	–
					HH/HV_PW	15.24	2.80	0.00001	0.92	1.08	1.89
					Standard deviation _3	10.29	3.92	0.01255	0.25	4.01	11.06
					Angular second moment_7	53.51	15.66	0.00159	0.58	1.74	3.16
					Inverses differences_5	−55.69	13.80	0.00027	0.47	2.14	5.11
					Homogeneity _9	28.40	10.67	0.01152	0.49	2.05	6.93
					Contrast _7	−22.83	7.57	0.00467	0.33	3.02	8.85
					Contrast _5	16.86	5.05	0.00197	0.22	4.51	8.49
					GLDV Mean _3	−6.60	3.06	0.03757	0.24	4.12	9.50

2007; Sarker et al. 2012) on land use/land cover mapping as well as biomass estimation, where multi-frequency, multi-polarization and multi-sensor data were used in the form of ratio or other combinations to improve accuracy levels using complementary information. Approximately 90 % accuracy (r^2) was obtained using the texture polarization ratio parameters. This is a very significant improvement and indicates that the ratio of dual polarization texture parameters is very effective for biomass estimation, and is adequately robust to represent high diversity and high biomass. This improvement is promising and agrees with the findings of previous researchers in principle who reported improvements using the ratio of multi-frequency and multi-polarization SAR compared with any single channel data (Ranson and Sun 1994; Shi and Dozier 1997; Foody et al. 1997; Sarker and Nichol 2011; Sarker et al. 2012). Although C-band SAR has previously been thought to be saturated at quite low levels (less than 50 t/ha) for biomass mapping (Le Toan et al. 1992; Imhoff 1995) this study confirms previous suggestions and findings (Luckman et al.1997; Castro et al. 2003; Oliver and Quegan 2004) that the recently available improved range of polarizations, incidence angles, and spatial resolutions as well as better image processing technique can vastly improve the sensitivity and overall performance of C-band SAR systems for forest mapping as well as biomass estimation.

References

Arevalo CBM, Volk TA, Bevilacqua E, Abrahamson L (2006) Development and validation of aboveground biomass estimations for four Salix clones in central New York. Biomass Bioenergy 31:1–12

Bortolot ZJ, Wynne RH (2005) Estimating forest biomass using small footprint LiDAR data: an individual tree-based approach that incorporates training data. J Photogrammetry Remote Sens 59:342–360

Brown S, Gillespie AJR, Lugo AE (1989) Biomass estimation methods for tropical forests with applications to forest inventory data. For Sci 35:881–902

Castro KL, Sanchez-Azofeifa GA, Rivard B (2003) Monitoring secondary tropical forest using space-borne data: implications for Central America. Int J Remote Sens 24:1853–1894

Champion I, Dubois-Fernandez P, Guyon D, Cottrel M (2008) Radar image texture as a function of forest stand age. Int J Remote Sens 29:1795–1800

Chen D, Stow DA, Gong P (2004) Examining the effect of spatial resolution and texture window size on classification accuracy: an urban environment case. Int J Remote Sens 25:2177–2192

Dell'Acqua F, Gamba P (2003) Texture-based characterization of urban environments on satellite SAR images. IEEE Trans Geosci Remote Sens 41:153–159

Dobson MC, Ulaby FT, Le Toan T, Beaudoin A, Kasischke Kasischke ES, Christensen NC (1992) Dependence of radar backscatter on conifer forest biomass. IEEE Trans Geosci Remote Sens 30:412–415

Dobson MC, Ulaby FT, Pierce LE, Sharik TL, Bergen KM, Kellndorfer J, Kendra JR, Li E, Lin YC, Nashashibi A, Sarabandi KL, Siqueira P (1995) Estimation of forest biomass characteristics in northern Michigan with SIR-C/X-SAR data. IEEE Trans Geosci Remote Sens 33:877–894

Foody GM, Green RM, Curran PJ, Lucas RM, Honzak M, Do Amaral I (1997) Observations on the relationship between SIR-C radar backscatter and the biomass of regenerating tropical forests. Int J Remote Sens 18:687–694

Foody GM, Boyd DS, Cutler MEJ (2003) Predictive relations of tropical forest biomass from Landsat TM data and their transferability between regions. Remote Sens Environ 85:463–474

Haralick RM, Shanmugam K, Dinstein I (1973) Texture features for image classification. IEEE Trans Syst, Man Cybern 3:610–621

Harrell PA, Kasischke ES, Bourgeau-Chavez LL, Haney EM, Christensen NL Jr (1997) Evaluation of approaches aboveground biomass in using SIR-C Data. Remote Sens Environ 59:223–233

Hyde P, Nelson R, Kimes D, Levine E (2007) Exploring LiDAR–RaDAR synergy—predicting aboveground biomass in a southwestern ponderosa pine forest using LiDAR, SAR and InSAR. Remote Sens Environ 106:28–38

Imhoff ML (1995) Radar backscatter and biomass saturation: ramification for global biomass inventory. IEEE Trans Geosci Remote Sens 33:511–518

Kasischke ES, Christensen NL, Bourgeau-Chavez LL (1995) Correlating radar backscatter with components of biomass in loblolly pine forests. IEEE Trans Geosci Remote Sens 33:643–659

Ketterings QM, Coe R, van Noordwijk M, Ambagau Y, Palm CA (2001) Reducing uncertainty in the use of allometric biomass equations for predicting above-ground tree biomass in mixed secondary forests. For Ecol Manage 146:199–209

Kuplich TM, Curran PJ, Atkinson PM (2005) Relating SAR image texture to the biomass of regenerating tropical forest. Int J Remote Sens 26:4829–4854

Le Toan T, Beaudoin A, Riom J, Guyon D (1992) Relating forest biomass to SAR data. IEEE Trans Geosci Remote Sens 30:403–411

Lu D (2006) The potential and challenge of remote sensing-based biomass estimation. Int J Remote Sens 27:1297–1328

Luckman A, Frery AC, Yanasse CCF, Groom GB (1997) Texture in airborne SAR imagery of tropical forest and its relationship to forest regeneration stage. Int J Remote Sens 18:1333–1349

Milne AK, Dong Y (2002) Vegetation mapping using JERS-1 SAR mosaic for northern Australia. Int J Remote Sens 23:1475–1486

Muukkonen P, Heiskanen J (2007) Biomass estimation over a large area based on standwise forest inventory data and ASTER and MODIS satellite data: A possibility to verify carbon inventories. Remote Sens Environ 107:617–622

Nelson R, Krabill W, Tonelli J (1988) Estimating forest biomass and volume using airborne laser data. Remote Sens Environ 24:247–267

Oliver CJ (1993) Optimum texture estimators for SAR clutter. J Phys D Appl Phys 26:1824–1835

Oliver C, Quegan S (2004) Understanding Synthetic Aperture Radar Images. SciTech Publication, Raleigh

Overman JPM, Witte HJL, Saldarriaga JG (1994) Evaluation of regression models for above-ground biomass determination in Amazon Rainforest. J Trop Ecol 10:207–218

Ranson KJ, Sun G (1994) Mapping biomass of a northern forest using multifrequency SAR data. IEEE Trans Geosci Remote Sens 32:388–396

Rauste Y, Hame T, Pulliainen J, Heiska K, Hallikainen M (1994) Radar-based forest biomass estimation. Int J Remote Sens 15:2797–2808

Rignot E, Way J, Williams C, Viereck L (1994) Radar estimates of aboveground biomass in boreal forests of interior Alaska. IEEE Trans Geosci Remote Sens 32:1117–1124

Rosenqvist Å, Milne A, Lucas R, Imhoff M, Dobson C (2003) A review of remote sensing technology in support of the Kyoto protocol. Environ Sci Policy 6:441–455

Sarker MLR, Nichol JE (2011) Improved forest biomass estimates using ALOS AVNIR-2 texture indices. Remote Sens Environ 115:968–977

Sarker MLR, Nichol JE, Ahmad B, Busu I, Rahman AA (2012) Improvement of forest biomass estimation using texture measurements of two-date dual polarization PALSAR data. ISPRS J Photogrammetry Remote Sens 69:146–166

Shi J, Dozier J (1997) Mapping seasonal snow with SIR-C/X-SAR in mountainous areas. Remote Sens Environ 59:294–307

Thenkabail PS, Stucky N, Griscom BW, Ashton MS, Diels J, Meer BVD, Enclona E (2004) Biomass estimations and carbon stock calculations in the oil palm plantations of African derived savannas using IKONOS data. Int J Remote Sens 25:5447–5472

Unser M (1986) Sum and difference histogram for texture classification. IEEE Trans Pattern Anal Mach Intell PAMI-8:118–125

Zheng D, Rademacher J, Chen J, Crow T, Bresee M, LE Moine J, Ryu S (2004) Estimating aboveground biomass using Landsat 7 ETM + data across a managed landscape in northern Wisconsin, USA. Remote Sens Environ 93:402–411

Reliability of a High Accuracy Image-Based System for 3D Modelling of the Medial Longitudinal Arch During Gait

Duaa Alshadli, Albert K. Chong, Kevin McDougall, Jasim Al-Baghdadi, Peter Milburn and Richard Newsham-West

Abstract The Medial Longitudinal Arch (MLA) is the largest arch of the foot and is regarded as the most important foot arch in clinical foot assessments due to its influence on lower limb function, foot stability and foot pain. Each foot is classified as either high arched, low arched or normally arched depending on the structure of the MLA. There are currently a number of techniques that are used to classify the foot in a static state based on measurements of the MLA. These static measurements are then used to predict the behaviour of the foot arch in a dynamic state. However, it is easy to identify limitations with these techniques as the shape of the MLA in a static state cannot predict the behaviour of the MLA during dynamic activities. Therefore, the aim of this chapter is to introduce a high accuracy 3D modelling system that has been developed to map the shape of the MLA during gait using high definition video camcorders. The objectives of the study were hence: (1) to determine whether changes can be detected along the MLA for different weight bearings during gait, (2) to test the accuracy and

D. Alshadli (✉) · A. K. Chong · K. McDougall · J. Al-Baghdadi
Faculty of Engineering and Surveying, University of Southern Queensland, Toowoomba, QLD, Australia
e-mail: duaa.alshadli@usq.edu.au

A. K. Chong
e-mail: chonga@usq.edu.au

K. McDougall
e-mail: kevin.mcdougall@usq.edu.au

J. Al-Baghdadi
e-mail: jasim76@gmail.com

P. Milburn · R. Newsham-West
School of Physiotherapy and Exercise Science, Griffith University, Gold Coast Campus, QLD, Australia
e-mail: p.milburn@griffith.edu.au

R. Newsham-West
e-mail: r.newsham-west@griffith.edu.au

A. Abdul Rahman et al. (eds.), *Developments in Multidimensional Spatial Data Models*, Lecture Notes in Geoinformation and Cartography,
DOI: 10.1007/978-3-642-36379-5_6, © Springer-Verlag Berlin Heidelberg 2013

reliability of the developed imaging system for creating dynamic 3D models of the foot arch and (3) to determine the quality and suitability of the 3D model. The results of the study show that changes can be detected along the MLA during gait with a level of accuracy of less than 0.4 mm when a 3D model of the foot is generated in PhotoModeler Scanner.

1 Introduction

In an attempt to identify the functionality of the human foot, a large number of studies have been dedicated to identifying the influence of foot morphology on foot function. The Medial Longitudinal Arch (MLA) in particular is identified as the most prominent structure of the foot which can be used to predict foot posture. Differences in foot morphology between individuals have been identified as a result of many factors including ethnicity (Igbigbi and Msamati 2002), type of foot wear (Pezzan et al. 2009), injury (Bennett et al. 2001; Bandholm et al. 2008) and gender (Nielsen et al. 2009). Razeghi and Batt (2002) provided a critical review of a myriad of methods used to examine the foot posture from MLA observations and summarised these techniques under the following four categories: (1) Visual inspection, (2) Anthropometric measurements, (3) Footprint measurements and (4) Radiographic measurements. The majority of these techniques are based on foot measurements in the standing weight-bearing position and do not take dynamic foot measurements into account.

Visual observations rely on the examiner's perception of visual aspects of the foot during weight bearing and are not useful for quantitative assessment of the MLA. This chapter will provide a quick review of some of the most commonly used quantitative assessment techniques.

1.1 Anthropometric Measurements

Anthropometric measurements involve the direct measurement of changes of a number of bony landmarks or soft tissue of the MLA both in standing and weight bearing stance. Two of the most extensively used anthropometric measurement techniques include the Navicular Drop and Navicular Drift which both use the Navicular bone as the reference point on the foot for studying the MLA during weight bearing. The Navicular Drop test (Fig. 1) provides a quantitative measure of the changes in the Navicular bone in the sagittal plane when weight-bearing on the foot is changed (Brody 1982), whereas the Navicular drift (Fig. 2) measures the changes in the Navicular bone in the medial–lateral plane during changes in weight bearing on the foot (McPoil et al. 2009).

Other common anthropometric tests include the Arch Height (Teyhen et al. 2009) which measures the changes in the dorsum height (vertical height at 50 %

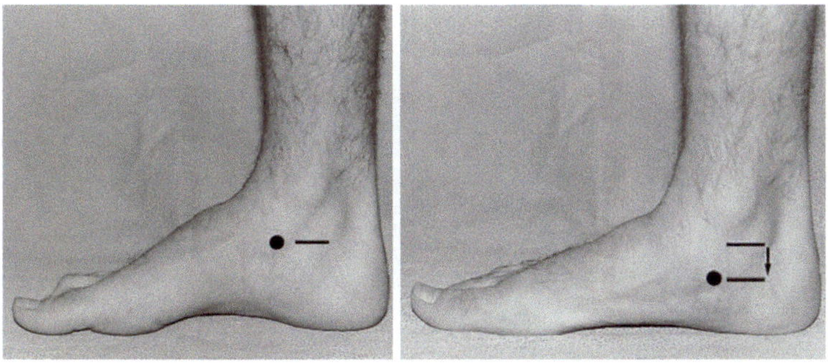

Fig. 1 Navicular drop as a result of change in weight bearing (Menz 1998)

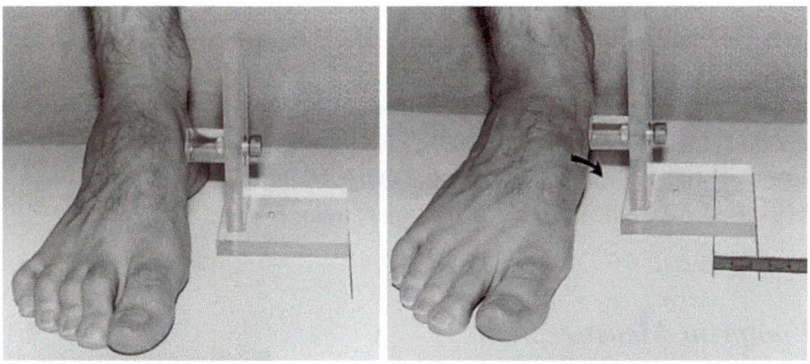

Fig. 2 Navicular drift as a result of change in weight bearing (Menz 1998)

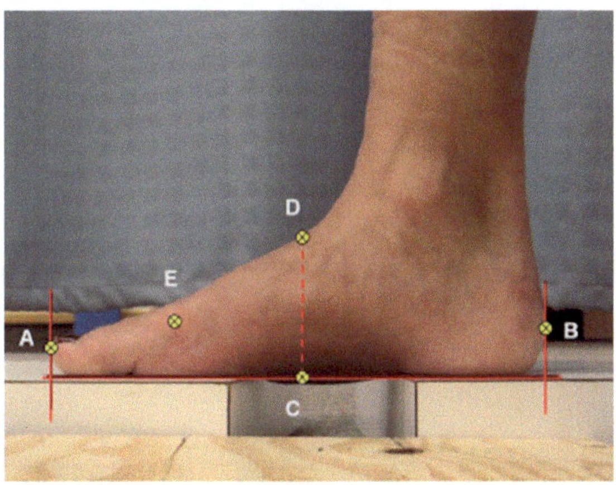

Fig. 3 Points used to calculate arch height and arch height index (Pohl and Farr 2010)

foot length) during weight bearing in a standing posture; and the Arch Height Index (AHI) which is calculated as a ratio of the of the dorsum height to the truncated foot length (distance between the heel and the first metatarsal head). Figure 3 is an illustration of the measurements used to determine the Arch height

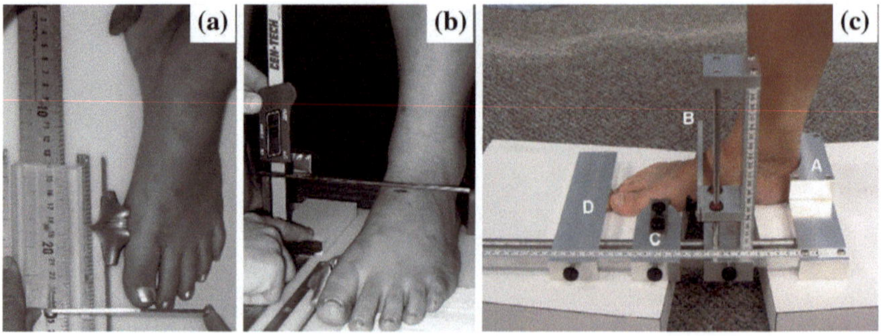

Fig. 4 **a** Rulers (McPoil et al. 2008). **b** Calliper (McPoil et al. 2008). **c** AHIMS (Pohl and Farr 2010)

and Arch Height Index where distance A–B is the total foot length, C–D is the dorsum height and the horizontal distance B–E is the truncated foot length.

The most commonly used equipment for anthropometric measurements include callipers and rulers (McPoil et al. 2008) or equipment adapted and constructed for the particular measurements such as the Arch Height Index measurement device (AHIMS) used by Pohl and Farr (2010) as shown in Fig. 4.

1.2 Footprint Measurements

Footprint measurements involve taking an imprint of the sole of the foot and the contact area with the surface of the imprint is used in the foot classification. A number of indices have been developed to quantify the footprint measurements and these are summarised in Fig. 5. The two most common indices are the Arch Index (AI) (Onodera et al. 2008) and the Footprint Index (Razeghi and Batt 2002).

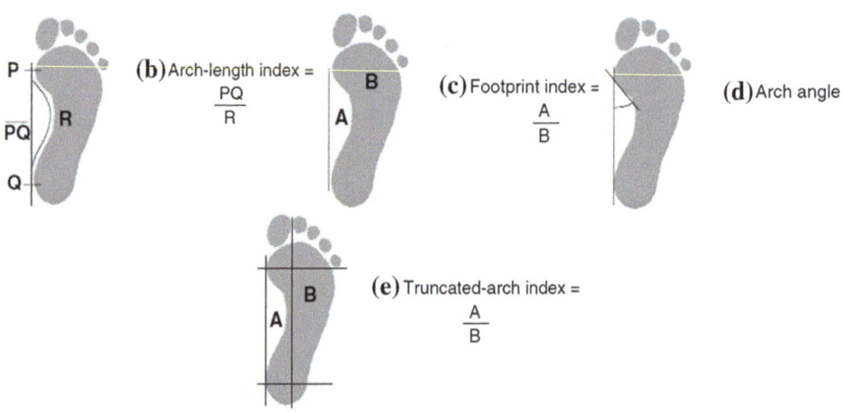

Fig. 5 Common footprint measurements (Menz 1998)

Footprint measurements are usually obtained with an ink mat, capacitive mat transducer (Chen et al. 2006) or a mirrored foot photo box (Mall et al. 2007).

1.3 Radiographic Measurements

Radiographic measurements of the foot have been identified as the gold standard in clinical examination of foot posture prediction as it can directly measure the locations of bony landmarks while the subject is standing in a full weight bearing position. Radiographs have become less feasible in recent years due to time consumption of data collection, expense and due to the negative effects on health from the radiographic exposure. Radiographs have been used to study the reliability of other foot posture measurement techniques and identify how the techniques correlate with each other (Menz and Munteanu 2005). The commonly derived measurements from radiographs which are used for comparisons with other static foot measurement techniques are shown in Fig. 6.

From the literature outlining the techniques used for predicting foot posture based on the measurements of the MLA, the main limitation identified is that all these techniques require the foot to be static during observations and the results are used to predict the behaviour of the MLA during dynamic activities such as gait or running. Measurements based on these techniques cannot be used readily to translate the shape of the MLA in a static state for predicting the behaviour of the MLA during dynamic activities. Another problem identified by Deng et al. (2010) was the variation in results between studies because of problems in identifying the exact amount of weight bearings used to calculate the changes in the foot arch during measurements. The aim of this study therefore is to resolve these issues by using a multi-sensor photogrammetric imaging system which can accurately predict the behaviour of a number of points on the MLA during gait. The objectives of the study are hence: (1) to determine whether changes can be detected along the MLA for different weight bearing during gait, (2) to test the accuracy and

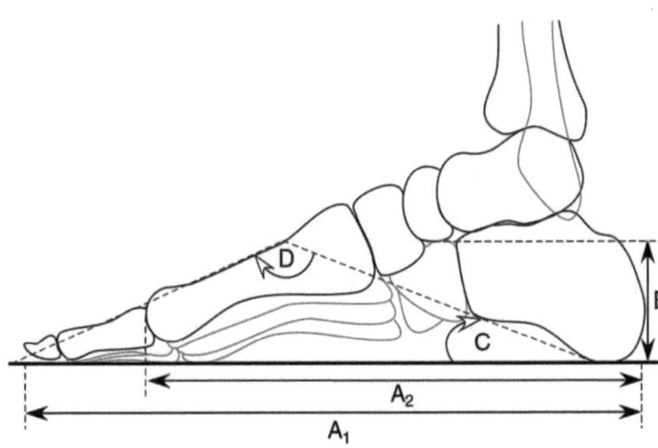

Fig. 6 Radiographic measurement used to measure the foot arch (Menz and Munteanu 2005)

reliability of the developed imaging system for creating dynamic 3D models of the foot arch and (3) to determine the quality and suitability of the 3D model.

The 3D foot model may then aid in understanding the behaviour of the MLA during gait as well as the relationship between the MLA and other foot features. The developed photogrammetric imaging system is comprised of High Definition (HD) camcorders that are able to image a moving object at 30 frames per second (fps), and a walking platform with a transparent force plate installed to allow for the simultaneous recording of the weight bearing under the foot during the gait cycle.

2 Image-Based Photogrammetric Modelling

To conduct image-based 3D measurements of an object, it is necessary to establish corresponding points between images (image matching) to establish a collinearity model for calculating the 3D coordinates of the points. Once the coordinates of points are calculated, it then becomes possible to calculate angular and distance measurements to a high level of accuracy. All the data needs to be relatively oriented between images with sufficient accuracy and scale to calculate distances accurately. For medical applications, imaging systems which are setup in overlapping stereo-pairs are the primary method used to capture 3D spatial data of targets (Chong et al. 2009).

For the stereo-pair imaging approach, a minimum of two imaging sensors must be used with 55–60 % overlap between them. Overlapping the image points using image matching allows for the determination of the 3D shape of the object (Mitchell and Newton 2002). Normally, a stereo-pair of images is essential to satisfy the photogrammetric algorithms such as the collinearity condition, coplanarity condition or epipolar geometry. Details of photogrammetric algorithms can be found in Luhmann et al. (2006). To improve the accuracy of calculated 3D points of an imaged object, more than one pair of stereo-images are used to produce a denser array of correspondences between images which when transformed through the camera model, can produce a Dense Surface Model (DSM) or a point cloud with X,Y and Z data.

Figure 7 illustrates a typical overlapping-pair photogrammetric configuration. For applications where high accuracy is essential, more than two stereo-pairs of imaging sensors are set up to image the target and provide more redundancy in the measurements and to make it easier to identify measurement blunders during image processing. Grobbelaar and Douglas (2007), Mutsvangwa et al. (2011) and Chong et al. (2009) used stereo-photogrammetry for 3D point generation with reported accuracies of less than 1 mm.

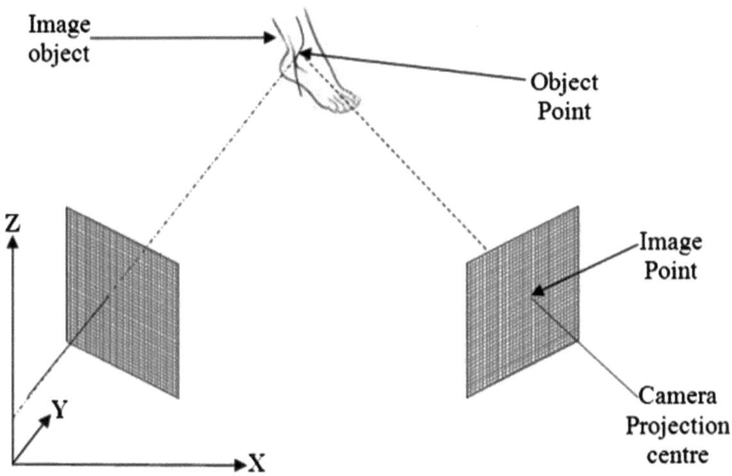

Fig. 7 Overlapping-pair photogrammetric configuration

3 Camera Calibrations

Due to the reduced cost and off-the-shelf availability, non-metric optical sensors are generally used for motion studies. However, with non-metric cameras the internal parameters can potentially affect measurement accuracy and must be accounted for through camera calibrations. Therefore, these cameras have deviations of the bundle of their imaging rays from a truly perspective model (Remondino 2004). The principal point offset (x_p, y_p), the principal distance (PD), the radial lens distortion parameters (K_1, K_2, K_3), decentring lens distortion parameters (P_1 and P_2) and affine distortion parameters (C_1 and C_2) must therefore be calculated through camera calibration prior to data capture using the bundle adjustment technique used by Remondino and El-Hakim (2006) and Wang and Clarke (2001). Collectively, these parameters are known as the imaging system's interior orientation parameters. The bundle adjustment process involves taking multiple convergent images of a pre-calibrated targeted grid from different angles and views and the imaged targets are used to create a resection-intersection of the bundle of rays based on the co-linearity condition (Eqs. 1 and 2). This condition states that the projection centre (O_i), the photo point (P_i) and the object point (P_j) fall in a straight as shown in Fig. 8.

The most effective calibration technique is self-calibration as it does not require high accuracy object-space coordinates. In addition to the interior orientation camera parameters, the bundle adjustment also solves for the camera exterior orientation parameters including the position of each camera. Equations 1 and 2 are used in photogrammetry self-calibration bundle adjustment (Wackrow et al. 2007).

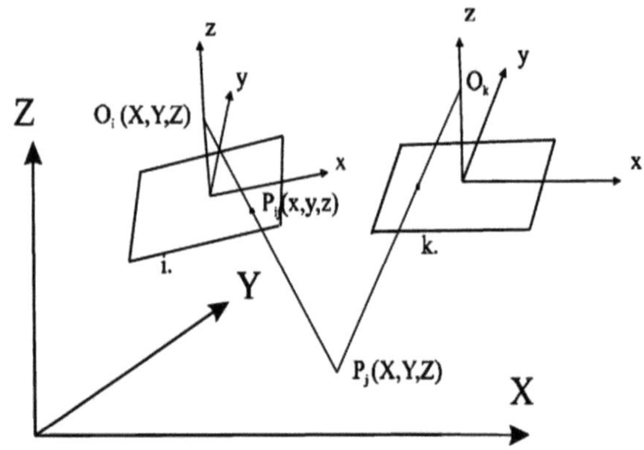

Fig. 8 The collinearity condition (Borlin 2002)

$$\Delta x' = \Delta x'_0 - \frac{x'}{PD}\Delta PD + K_1 x' r'^2 + K_2 x' r'^4 + K_3 x' r'^6 + P_1\left(r'^2 + 2x'^2\right) + 2P_2 x' y' - C_1 x' + C_2 y' \quad (1)$$

$$\Delta y' = \Delta y'_0 - \frac{y'}{PD}\Delta PD + K_1 y' r'^2 + K_2 y' r'^4 + K_3 y' r'^6 + 2P_1 x' y' + P_2\left(r'^2 + 2y'^2\right) + C_2 x' \quad (2)$$

where

$\Delta x'$, $\Delta y'$	Axis-related correction values for imaging errors,
$\Delta x'_0, \Delta y'_0$	Small corrections for perspective centre (X_p, and Y_p),
ΔPD	Small correction for principal distance,
K_1, K_2, K_3	Lens distortion parameters,
P_1, P_2	Lens decentring parameters,
C_1, C_2	Affinity and shear parameters, and
r'	Radial distance.

4 Equipment and Software

4.1 Equipment

To be able to model the human foot arch during gait or any other dynamic activity, a video image capture device needs to be employed. For the purpose of our study, a total of four High Definition (HD) JVC Everio GZ-HD500 camcorders with a focal length varying from 3.0 to 60 mm were utilised. During the data capture, the subjects walked on an elevated platform with a transparent force plate installed to determine the variations in weight bearing during different segments of the gait cycle. As the images and the force plate data needed to be synchronised, a generator locking (gen-lock) device was developed to allow for a flashing LED light to be displayed as soon as the foot touched the force plate. The gen-lock device capture a set of hard-wired or remotely controlled signals while simultaneously processing another set (video). This ensured that if more than one video system was used, the same signal at the same instant of time was received by all the video footages.

4.1.1 Software

After the video clips were recorded, images were frame-grabbed using Tripard HD video converter software (version 6.1.12) and the format was converted from MPG to JPG image format using Virtual Dub software. To record the force plate data, the NetForce AMTI software was run on a laptop to continuously record the force data during video imaging. For each individual camera calibration and the calculation of the 3D points on the foot, the software Australis (version 6.06) was used. Australis was selected as the ‘gold standard’ of measurement accuracy due to its proven low RMS values for individual measurements of landmarks (Al-Baghdadi et al. 2011). PhotoModeler Scanner software (version 6) was chosen to create a graphical 3D representation of the foot and the measurements obtained in PhotoModeler were compared to those obtained from Australis.

5 Methods

5.1 Video Camcorder Calibration

5.1.1 PhotoModeler Scanner

For the purpose of camera calibration, each camcorder was set up at a distance of 1 m from a calibration test field obtained from PhotoModeler Scanner as shown in Fig. 9. The test field contains 100 coded targets with a configuration which can be

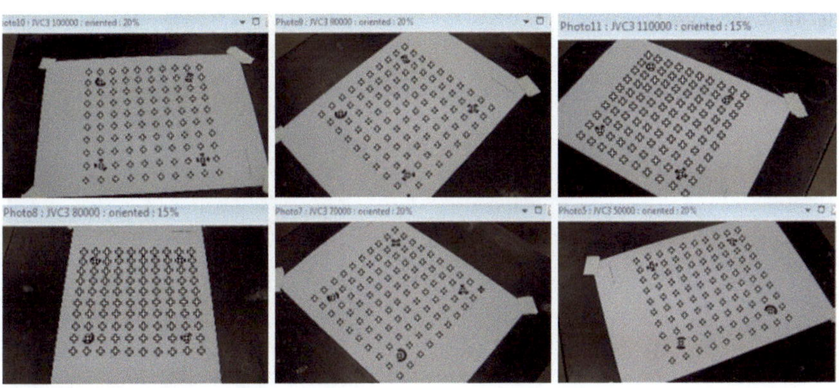

Fig. 9 The test-field supplied by PhotoModeler Scanner

automatically recognised in PhotoModeler Scanner. Eight convergent video recordings were collected (four viewing the calibration board vertical and four rotated to 90°) and still images were extracted from the video clips using Tripard HD video converter and VirtualDub. Once the images were extracted, they were imported into PhotoModeler Scanner and automatically processed using the camera calibration tool to obtain each camera's interior orientation parameters. Each camera's intrinsic parameters were then stored into a library for each individual camera which was later used to assign the project images to the subsequent camera from the camera library.

5.1.2 Australis

Camera calibrations using Australis were conducted in a manner similar to the calibrations completed in PhotoModeler. The only difference was in the calibration targets which consisted of retro-reflective targets placed on a calibration board as shown in Fig. 10.

Once all eight convergent video frames were recorded and the still images converted, an initial focal length distance of 3.0 mm was entered and the points were digitised manually in Australis and following the bundle adjustment, the calibration results were determined (Table 1).

For the purpose of this study, only the PD, the principal point offsets (x_p, y_p) and the radial lens parameter K_1 were determined to be significant in affecting the accuracy for this particular project. K_2 and K_3, the decentring lens distortion parameters (P_1, P_2) and the affine distortion parameters (C_1, C_2) were excluded due to their small error contribution to the real-world coordinates. An example of a radial distortion plot is shown in Fig. 11.

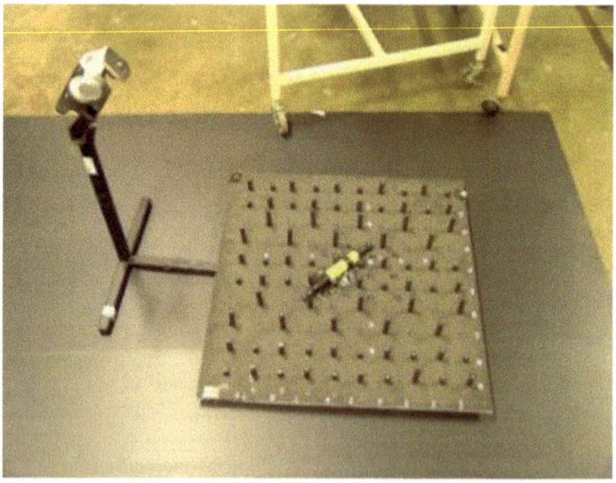

Fig. 10 Calibration board used to calibrate cameras in Australis

Table 1 Results of the camera calibration obtained in Australis

Camera ID	PD (mm)	xp (mm)	yp (mm)	K1(mm)	K2 (mm)	K3 (mm)
JVC 1	3.1611	−0.0323	−0.0173	0.0028	−0.00088	−0.00040
JVC 2	3.4502	−0.0433	−0.0021	0.0023	−0.00072	−0.00018
JVC 3	3.1779	0.0097	0.0075	0.0011	−0.00027	−0.00018
JVC 4	3.4498	−0.0163	−0.0113	0.0024	−0.00015	−0.00005

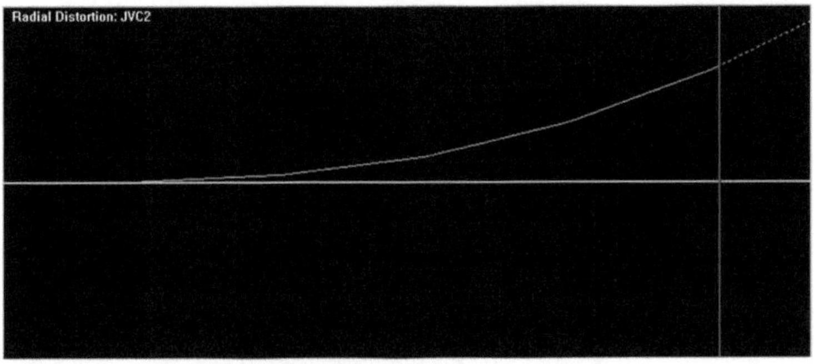

Fig. 11 Example of Radial lens distortion plot

5.2 Imaging During Gait

An elevated platform (1 m above the ground) was designed and constructed in such a way to allow images to be recorded from the medial view of the foot while constantly recording the changes in weight bearing under the foot during gait. For this, a transparent force plate was installed on the platform where the subject's foot landed during gait. The setup is shown in Figs. 12 and 13.

Each subject's foot of interest was painted as a previous study found that painting the foot created improved texture and contrast and hence yielded better results when processed in PhotoModeler (Alshadli et al. 2011). Each subject was then instructed to walk across the platform at a self-selected pace and the moment at which the subject's foot landed on the force plate, a gen-lock system was activated and a flashing light was triggered as shown in Figs. 13 and 14. A number of control targets were also added around the force plate to allow for relative orientation of images in PhotoModeler. Artificial lighting had to be used during imaging to eliminate any effects of shadows on the imaged object.

5.2.1 Computation of Point Coordinates in Australis

In order to calculate the coordinates of the points marked on the medial side of the foot, two stereo-pairs of images were relatively oriented for each step where the weight bearing changed. Weight bearings under foot of 10, 50, 90 and 100 % were

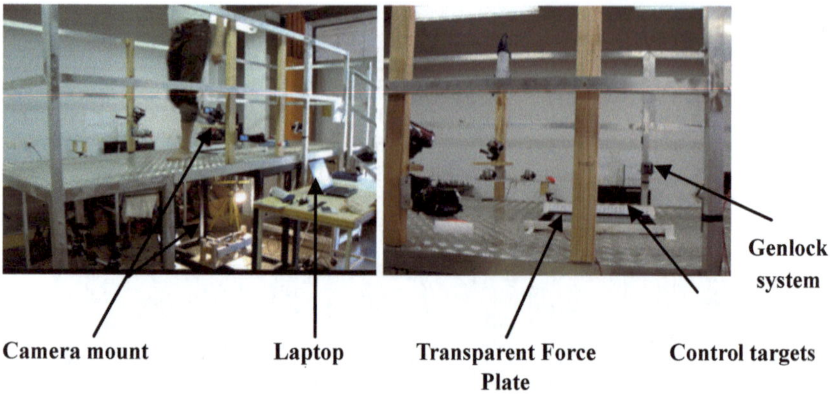

Fig. 12 Elevated platform with imaging system

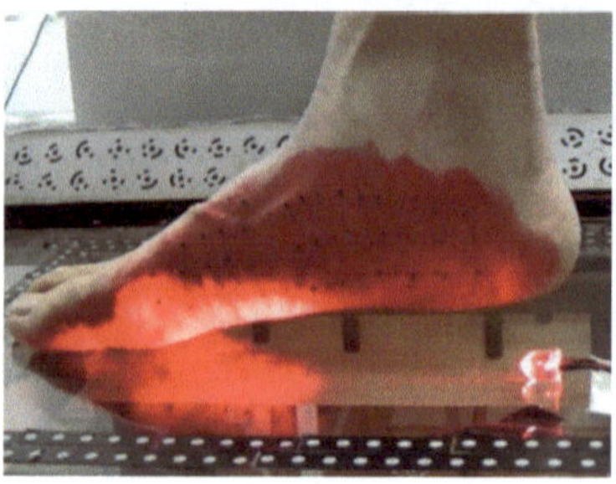

Fig. 13 The moment the gen-lock system is activated and the LED light becomes visible

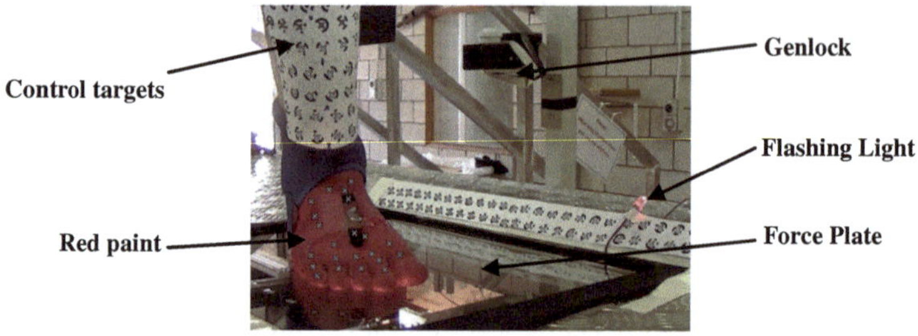

Fig. 14 Camera synchronisation

chosen for the analysis. The camcorders were synchronised with the force plate at the instant of the selected weight bearing by using the flashing light and correlating the time of the force plate with the subsequent frame number in Virtual Dub. Figure 15 shows the location of the image selected on the force graph.

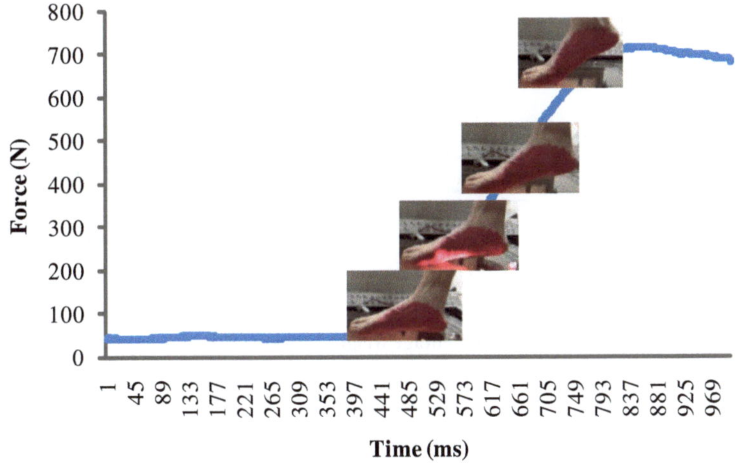

Fig. 15 Correlation between pose and weight bearing

After the stereo-pairs of images were relatively oriented, the bundle adjustment was used to calculate the 3D coordinates for all the marked points.

5.2.2 Computation of the Dense Surface Model

The same images used in Australis were also imported into PhotoModeler. Each image was then assigned to the camera calibration parameters which were previously stored in the camera library as discussed in Sect. 5.1.1. Image matching was then conducted and all the images were relatively oriented before they were processed to create a Dense Surface Model (DSM) of the foot.

6 Results and Discussion

From the 48 points marked on the medial side of the foot, six points were selected in Australis at random along the medial side of the foot for data comparisons during the different percentages of weight bearing. The X, Y and Z coordinates were plotted separately on a graph for each point and the results for an average of two sets of data are displayed in Fig. 16a, b, and c for coordinates X, Y and Z respectively.

All six points that were selected displayed a similar trend during the different weight bearing percentages for the average of the two sets of data even though the points were randomly selected across the entire medial side of the foot. From this, it can be concluded that a few points will be sufficient to track changes in the medial side of the foot during the different weight bearings of gait. It can also be

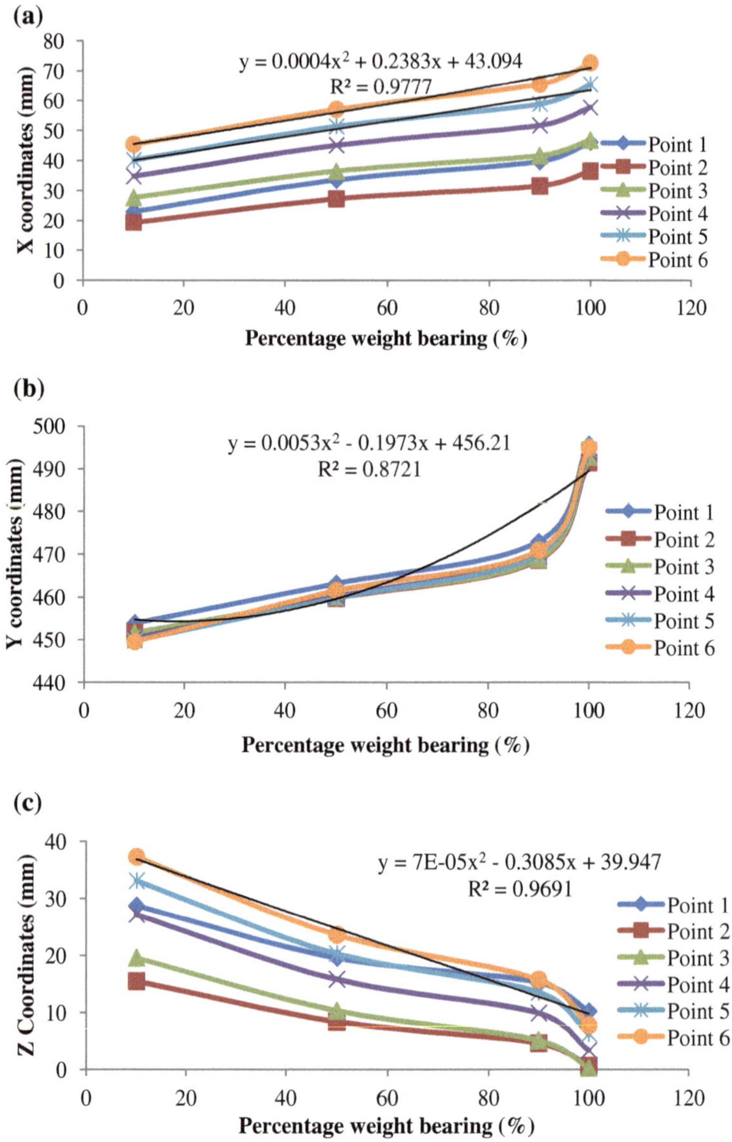

Fig. 16 Relationship between coordinates and weight bearings during gait

shown from the graphs that the correlation between the coordinates and the weight bearings can be described using polynomial equations.

From Fig. 17, it can be seen that the four camera configuration (2 stereo-pairs) was insufficient to provide a complete 3D model of the foot due to obstructions between the camera views. However, the common view area between the cameras

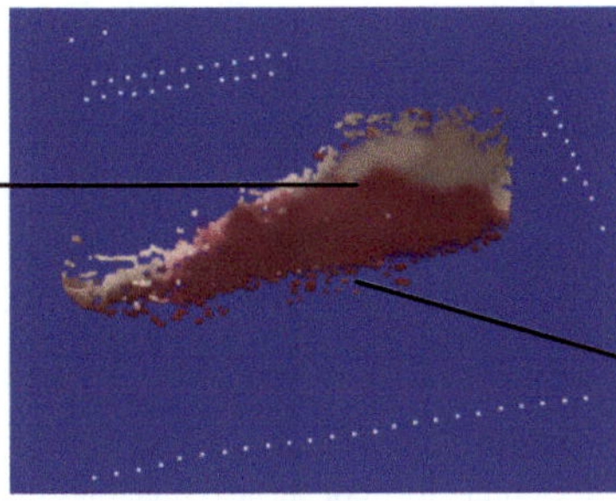

Fig. 17 3D model generated in PhotoModeler

Table 2 Accuracy assessment of points in PhotoModeler

Points	Distance in Australis (mm)	Distance in PM (mm)	Difference (mm)
1–2	10.802	10.695	0.107
3–4	9.066	8.718	0.347
5–6	12.824	13.148	0.324

provided a Dense Surface Model with sufficient data points. In our current study, this included our area of interest which represents the MLA.

To validate the accuracy of the 3D model, three distances between points on the 3D model were compared to the same distance between points in Australis and the results are summarised in Table 2. Australis was selected as the 'gold standard' for measurement accuracy due to its low RMS measurement values (Al-Baghdadi et al. 2011). The comparison of distances between points provided an indication of the accuracy of the measurements in PhotoModeler with the largest difference being less than 0.4 mm (Table 2).

7 Conclusion

The foot medial longitudinal arch is used to predict foot posture and identify foot functionality. However, the current foot posture prediction techniques rely on static measurements of the foot MLA to predict the dynamic behaviour of the foot. The current research developed a novel technique which can be used to model the MLA during gait while simultaneously measuring the dynamic weight bearing under the foot during gait. From the results obtained in the study, it is concluded that the developed technique can measure and correlate the dynamic weight bearing during gait to the coordinates of points on the medial side of the foot. In addition, the 3D model created in PhotoModeler provided sufficient dense surface data to model the area of the MLA. However, if a full view of the foot is required to be able to relate the behaviour of the MLA with foot functionality, then more camcorders need to be used. In the future, the technique can be developed further to predict the behaviour of the MLA during faster movements such as running by using higher speed video cameras.

References

Al-Baghdadi J, Chong AK, Alshadli D, McDougall K, Milburn P, Newsham-West R (2011) A dense surface modelling technique for foot surface imaging. In: Proceedings of the surveying and spatial sciences biennial conference, Wellington

Alshadli D, Chong AK, McDougall K, Al-Baghdadi J (2011) Skin texture-enhancement for automated 3D point cloud generation in charcot-marie-tooth disease application. In: The tenth international conference on electronic measurement and instruments, pp 6–10

Bandholm T, Boysen L, Haugaard S, Zebis MK, Bencke J (2008) Foot medial longitudinal-arch deformation during quiet standing and gait in subjects with medial tibial stress syndrome. J Foot Ankle Surg 47:89–95

Bennett J, Reinking M, Pluemer B, Pentel A, Seaton M, Killian C (2001) Factors contributing to the development of medial tibial stress syndrome in high school runners. J Orthop Sports Phys Ther 31:504–510

Borlin N (2002) Comparison of resection-intersection algorithms and projection geometries in Radiostereometry. ISPRS J Photogram Remote Sens 56:396–400

Brody D (1982) Techniques in the evaluation and treatment of the injured runner. Orthop Clin N Am 13:541–558

Chen C-H, Huang M-H, Chen T-W, Weng M-C, Lee C-L, Wang G-J (2006) The correlation between selected measurements from footprint and radiograph of flatfoot. Arch Phys Med Rehabil 87:235–240

Chong AK, Milburn P, Newsham-West R, Voert M (2009) High-accuracy photogrammetric technique for human spine measurement. Photogram Rec 24:264–279

Deng J, Joseph R, Wong CK (2010) Reliability and validity of the sit-to-stand navicular drop test: do static measures of navicular height relate to the dynamic navicular motion during gait? J Stud Phys Ther Res 2:21–28

Grobbelaar R, Douglas TS (2007) Stereo image matching for facial feature measurement to aid in fetal alcohol syndrome screening. Med Eng Phys 29:459–464

Igbigbi PS, Msamati BC (2002) The footprint ratio as a predictor of pes planus: a study of indigenous Malawians. J Foot Ankle Surg 41:394–397

Luhmann T, Robson S, Kyle S, Harley I (2006) Close range photogrammetry: principles, techniques and applications. Whittles Publishing, Caithness

Mall NA, Hardaker WM, Nunley JA, Queen RM (2007) The reliability and reproducibility of foot type measurements using a mirrored foot photo box and digital photography compared to caliper measurements. J Biomech 40:1171–1176

Mcpoil TG, Cornwall MW, Vicenzino B, Teyhen DS, Molloy JM, Christie DS, Collins N (2008) Effect of using truncated versus total foot length to calculate the arch height ratio. The Foot 18:220–227

Mcpoil TG, Vicenzino B, Cornwall MW, Collins N, Warren M (2009) Reliability and normative values for the foot mobility magnitude: a composite measure of vertical and medial-lateral mobility of the midfoot. J Foot Ankle Res 2:12

Menz HB (1998) Alternative techniques for the clinical assessment of foot pronation. J Am Podiatr Med Assoc 88:119–129

Menz HB, Munteanu SE (2005) Validity of 3 clinical techniques for the measurement of static foot posture in older people. J Orthop Sports Phys Ther 35:479–486

Mitchell HL, Newton I (2002) Medical photogrammetric measurement: overview and prospects. ISPRS J Photogram Remote Sens 56:286–294

Mutsvangwa TEM, Veeraragoo M, Douglas TS (2011) Precision assessment of stereo-photogrammetrically derived facial landmarks in infants. Ann Anat 193:00–105

Nielsen RG, Rathleff M, Simonsen O, Langberg H (2009) Determination of normal values for navicular drop during walking: a new model correcting for foot length and gender. J Foot Ankle Res 2:7

Onodera A, Sacco ICN, Morioka E, Souza P, DeśA M, Amadio A (2008) What is the best method for child longitudinal plantar arch assessment and when does arch maturation occur? Foot 18:142–149

Pezzan PAO, Sacco ICN, João SMA (2009) Foot posture and classification of the plantar arch among adolescent wearers and non-wearers of high-heeled shoes. Rev Bras Fisioter, São Carlos 13:398–404

Pohl MB, Farr L (2010) A comparision of foot arch measurement reliability using both digital photography and calliper methods. J Foot Ankle Res 3:14

Razeghi M, Batt M (2002) Foot type classification: a critical review of current methods. Gait Posture 15:282–291

Remondino F (2004) 3-D reconstruction of static human body shape from image sequence. Comput Vis Underst 93:65–85

Remondino F, El-Hakim S (2006) Image based 3D modelling: a review. Photogram Rec 21:269–291

Teyhen DS, Stoltenberg BE, Collinsworth KM, Giesel CL, Williams DG, Kardouni CH, Molloy JM, Goffar SL, Christie DS, Mcpoil T (2009) Dynamic plantar pressure parameters associated with static arch height index during gait. Clin Biomech 24:391–396

Wackrow R, Chandler J, Bryan P (2007) Geometric consistency and stability of consumer-grade digital cameras for accurate spatial measuement. Photogram Rec 22:121–134

Wang X, Clarke T (2001) Separate adjustment in close-range photogrammetry. ISPRS J Photogram Remote Sens 55:289–298

WorldView-2 Satellite Imagery and Airborne LiDAR Data for Object-Based Forest Species Classification in a Cool Temperate Rainforest Environment

Zhenyu Zhang and Xiaoye Liu

Abstract High resolution spatial data including airborne LiDAR data and newly available WorldView-2 satellite imagery provide opportunities to develop new efficient ways of solving conventional problems in forestry. Those responsible for monitoring forest changes over time relevant to timber harvesting and native forest conservation realize the potential for improved documentation from using such data. Proper use of high spatial resolution data with object-based image analysis approach and nonparametric classification method such as decision trees offers an alternative to aerial photograph interpretation in support of forest classification and mapping. This study explored ways of processing airborne LiDAR data and WorldView-2 satellite imagery for object-based forest species classification using decision trees in the Strzelecki Ranges, one of the four major Victorian areas of cool temperate rainforest in Australia. The effectiveness of variables derived from different data sets, in particular, the four new bands of WorldView-2 imagery was examined. The results showed that using LiDAR data alone or four conventional bands only, the overall accuracies achieved were 61.39 and 61.42 % respectively, but the overall accuracy increased to 82.35 % when all eight bands and the LiDAR data were used. This study demonstrated that the integration of airborne LiDAR and eight WorldView-2 bands significantly improved the classification accuracy.

Keywords WorldView-2 · LiDAR · Object-based image analysis · Forest classification · Decision tree · Cool temperate rainforest

Z. Zhang (✉) · X. Liu
Australian Centre for Sustainable Catchments, and Faculty of Engineering and Surveying, University of Southern Queensland, Toowoomba, QLD 4350, Australia
e-mail: zhenyu.zhang@usq.edu.au

Z. Zhang
Centre for GIS, School of Geography and Environmental Science, Monash University, Clayton, VIC 3800, Australia

A. Abdul Rahman et al. (eds.), *Developments in Multidimensional Spatial Data Models*, Lecture Notes in Geoinformation and Cartography,
DOI: 10.1007/978-3-642-36379-5_7,

1 Introduction

Forest classification and mapping is part of decision support for sustainable forest management and native forest conservation (Franklin 2001). The traditional methods for forest classification were based on an interpretation of large scale aerial photographs and/or a field survey. These methods are labour intensive and time consuming. Remotely sensed imagery has a wide range of applications in forestry. However, the application of satellite image data for forest classification and mapping with moderate spatial resolution image data such as images from the Landsat satellite has produced limited results, as the spatial resolution of the sensors is too coarse to delineate small forest patches. Newly available high spatial resolution imagery from several commercial satellites such as Ikonos, QuickBird, and more recently WorldView-2 has the potential to map forests in much more detail than ever before. However, high spatial resolution imagery poses challenges in image classification. In high spatial resolution images, individual pixels are typically smaller than the objects of interest such as forest stands and tree crowns. Thus, traditional pixel-based classification based solely on spectral data may not work successfully and can result in salt-and-pepper noise in the classification results due to higher spectral variance within the objects of interest (Ke et al. 2010; Benz et al. 2004). To minimize the effect of high spectral variance, object-based image analysis (OBIA) was introduced to remote sensing communities (Hay et al. 1995; St-Onge and Cavayas 1997; Pekkarinen 2002; Benz et al. 2004; Baatz and Schäpe 2000).

The OBIA approach segments the image into homogeneous groups of pixels (i.e., objects) prior to the classification of such objects. Image segmentation emerged as early as the 1970s and has been extensively used for a range of applications. Within the field of remote sensing, OBIA has shown a rapid development and applications in the last 10 years, mainly as a consequence of the increased spatial resolution of the newer remote sensors (Arroyo et al. 2010). The basic units of object-based classification are image objects or segments. Image objects are generated using an image segmentation procedure, which partitions an image into non-intersecting regions. Each object is composed of spatially adjacent pixels clustered according to homogeneity in response to selected segmentation criteria (Ke et al. 2010; Benz et al. 2004). One of the advantages of object-based image analysis is the additional information that can be derived from image objects. Image objects containing multi-pixel groups enable the calculation of aggregative statistics, such as mean and standard deviation of digital numbers of pixels within an object. In addition to spectral-based information, information based on object size, shape, and context can also be calculated for image analysis (Chubey et al. 2006; Blaschke 2010).

Although remotely sensed data have been widely explored for forest applications, passive remote sensing techniques that sense naturally available energy (Lillesand et al. 2008) are limited in their ability to capture forest structural complexity, particularly in uneven-aged, mixed species forests with multiple canopy layers

(Lovell et al. 2003). Generally, these techniques are only able to provide information on horizontal (two-dimensional) forest extent (Danson et al. 2009; Lim et al. 2003; Riaño et al. 2003). The vertical forest structure (or the interior of the canopy and understorey vegetation) cannot be mapped using these passive remote sensing techniques (Popescu et al. 2002; Scanlan et al. 2010; Morsdorf et al. 2010). A sensor that can see through the canopy and capture the three-dimensional structure of the forest is needed. Fortunately, it has been shown that active remote sensing techniques via airborne LiDAR (light detection and ranging) with canopy penetration capability can yields such high density sampling that detailed description of the forest structure in three-dimensions can be obtained (Popescu et al. 2002; Evans et al. 2006; Zhang and Liu 2012). For forest applications, multiple return LiDAR data can provide detailed information not only from the top of the canopy, but also through the interior of the canopy including the mid-storey and the lower storey, and the ground (Zhang et al. 2011). It offers great potential for quantifying forest structure in support of a wide range of applications such as forest inventory data extraction, forest species identification and classification.

The LiDAR data have been shown to measure forest structure characteristics accurately in a variety of forest types. However, the LiDAR data alone cannot provide all the information about forest composition (Erdody and Moskal 2010). The LiDAR data may not discern tree species very well, especially for those with similar canopy structure and composition. It has been recognised that integration of high spatial resolution imagery and LiDAR data can improve forest classification since the information on the vertical forest structure from the LiDAR data complements the spectral information from the optical imagery (Ke et al. 2010; Holmgren et al. 2008).

With the increasing availability of remotely sensed data from multiple sources and the increasing use of OBIA in forest applications, classification approaches capable of processing large input datasets are required to take full advantage of the rich set of information derived from OBIA (Chubey et al. 2006). Furthermore, the problems associated with satisfying the assumptions that underlie traditional classification methods have also driven research into nonparametric alternatives such as decision trees. Decision trees have advantages over traditional classification methods used in remote sensing such as maximum likelihood classification (MLC). Decision trees can also handle a wide variety of input data. Unlike conventional statistical classifiers, which use all available variables simultaneously and make a single membership decision for each object, the decision tree uses a multi-stage or sequential approach to the problem of classification (Pal and Mather 2003).

Decision trees have been applied to the analysis of remotely sensed data in several contexts including land cover classification (Pal and Mather 2003; Hansen et al. 1996; Lawrence and Wright 2001; Lawrence et al. 2004; Friedl and Brodley 1997; Borak and Strahler 1999) and vegetation mapping (Brown de Colstoun et al. 2003; Michaelsen et al. 1994; Potapov et al. 2008; Yu et al. 2006; Mallinis et al. 2008). There has been increasing interest in the use of decision trees in object-based classification because of the increasing availability of high spatial resolution imagery and airborne LiDAR data. For example, Ke et al. (2010) applied decision trees to the analysis of variables derived from LiDAR and high spatial resolution QuickBird multispectral data for object-based forest species classification in central New York State. Sasaki et al. (2010) used decision trees in the OBIA of airborne LiDAR and high spatial resolution imagery data to classify land cover types and dominant tree species in Osaka, Japan.

Increased spatial resolution also provides opportunities to develop new efficient ways of solving conventional problems in forestry. Those responsible for monitoring forest changes over time relevant to timber harvesting and native forest conservation see the potential for improved documentation from using such data. In Victoria, Australia, Ecological Vegetation Classes (EVCs) are the basic regional-scale mapping unit describing the spatial extent of native vegetation species for forest ecosystem assessments, biodiversity planning and conservation management. If newly available high resolution spatial data including airborne LiDAR and WorldView-2 satellite imagery can be used to accurately discriminate forest structural classes, these new technologies may offer an alternative to aerial photograph interpretation so as to enable more regular updating of EVC mapping polygons in support of native forest conservation in Victoria.

The WorldView-2, launched on 8 October 2009, is the first commercial satellite to provide eight multispectral bands with 1.8 m spatial resolution (at nadir), with four conventional bands similar to those provided by QuickBird: blue, green, red, and near infra-red 1 (NIR1), and four new bands: coastal blue, yellow, red-edge, and NIR2, and 0.46 m spatial resolution (at nadir) panchromatic imagery (DigitalGlobe 2009). The WorldView-2 satellite imagery is so new that few studies have been carried out to test its effectiveness in forest applications (Ozdemir and Karnieli 2011; Novichikhin 2011). The suitability of integration of airborne LiDAR data and WorldView-2 satellite imagery for object-based forest classification, particularly in Australian cool temperate rainforest environment has hitherto remained untested. The overall objective of this study was to explore the applicability of airborne LiDAR data and WorldView-2 satellite imagery data for object-based forest species classification in the Strzelecki Ranges, one of the four major Victorian areas of cool temperate rainforest. Specific objectives included the test of the effectiveness of LiDAR-derived variables based on the height distribution of LiDAR points within each of square columns, the contribution of new WorldView-2 image bands to object-based forest classification.

2 Study Area and Materials

2.1 Study Area

The study area is in the Strzelecki Ranges, southeast Victoria, Australia (Fig. 1). The Strzelecki Ranges, are an isolated series of mountains with an area of 3,418.6 km^2 (WGCMA 2009), in the southern section of the Gippsland region that

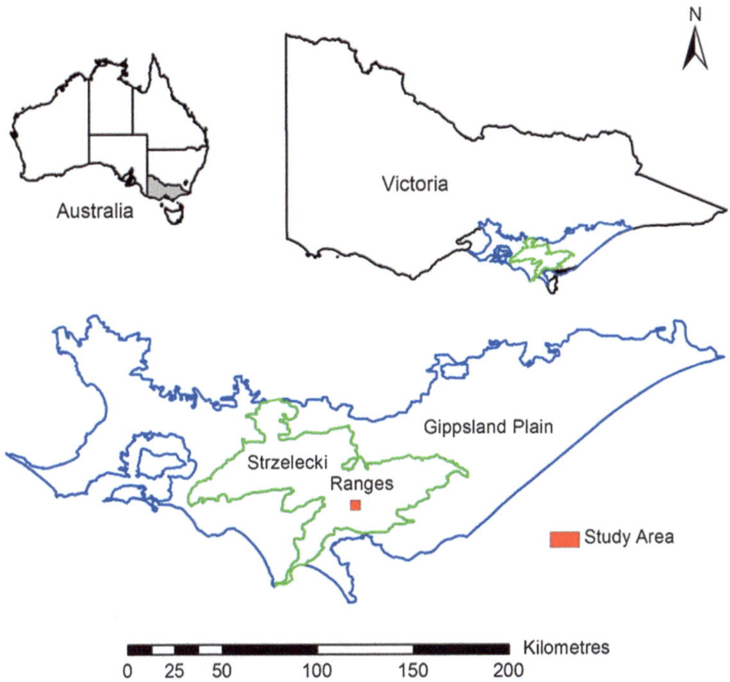

Fig. 1 Study area

are surrounded by the Gippsland Plain. Prior to European settlement, the Strzelecki Ranges were densely vegetated by wet forest (also referred to as wet sclerophyll forest) and cool temperate rainforest. Wet forest is most commonly dominated by Mountain Ash (*Eucalyptus regnans*) (Davies et al. 2002), characterised by a tall eucalypt overstorey, a broad-leaved shrubby understorey and a moist, shaded, fern-rich ground layer that is usually dominated by tree-ferns (DSE 2011). In eucalypt-free areas, Silver Wattle (*Acacia dealbata*) may be locally dominant (Davies et al. 2002). Cool temperate rainforest is defined as a closed, non-eucalypt forest, which occurs in high rainfall areas and within wet forest areas which have not been exposed to fire (Adam 1992). Myrtle Beech (*Nothofagus cunninghamii*) is the dominant species of cool temperate rainforest in the study area. The understorey is characterised by tree ferns and a rich epiphytic flora. The ground layer is dominated by a diversity of ground ferns such as Mother Shield-fern, Hard Water-fern, and Leathery Shield-fern (DSE 2011; Peel 1999). These forests have experienced widespread land clearing since European settlement. Subsequent agricultural abandonment and a significant fire history have resulted in severe landscape disturbance in the Strzelecki Ranges (Noble 1978; Legg 1986). There was extensive regeneration of eucalypt forest following catastrophic wildfires in 1939 and 1944. The landscape has undergone further significant changes with the establishment of large scale plantations in the area over the last half of a century. Plantations include softwood Radiata Pine and hardwood eucalypts. Currently, the mosaic of land covers bordering cool temperate rainforest in the Eastern Strzeleckis is the legacy of the different land use histories involving both natural and human disturbances, and so a very complex forest structure prevails in the remnant patches of cool temperate rainforest and adjacent forests including wet sclerophyll and plantation forests. This study focuses on an area with cool temperate rainforest distribution in the Eastern Strzeleckis, which covers an area of 3.10 km^2 with elevations ranging between 300 and 458 m, shown in Fig. 1. There are five forest types in the study area: cool temperate rainforest (RF), naturally regenerated Mountain Ash (MA), mixed forest (MF) consisting of overstorey Mountain Ash and understorey Myrtle Beech, Silver Wattle (SW), and hardwood plantation (PT) dominated by Shining Gum (*Eucalyptus nitens*).

2.2 Data

LiDAR data covering the whole area of the Strzelecki Ranges were collected using an Optech ALTM Gemini LiDAR system at a flying height of 1,100 m between 11 and 23 October 2009. The laser pulse repetition frequency is 70 kHz. The laser scanner was configured to record up to four returns per laser pulse. The average point density was 4 points per square metre, and the laser footprint diameter (the diameter of the laser pulse sampling area on the target) was 0.3 m. The LiDAR data used for this project were documented as 0.20 m for vertical accuracy and 0.25 m for horizontal accuracy. The LiDAR data were classified into ground and non-ground points by the vendor and were delivered in binary LAS 1.2 file format.

WorldView-2 satellite imagery over the study area was acquired on 3 December 2010. Each of eight multispectral bands is narrowly focused on a particular range of the electromagnetic spectrum that is sensitive to a particular feature on the ground. WorldView-2 is the first commercial satellite to provide a red-edge band as part of its 8-band multispectral capabilities. Access to red-edge data is enabling novel remote sensing applications in forest classification that depends on the detection of subtle changes in plant communities (DigitalGlobe 2009). Images were ortho-rectified by the vendor and delivered at spatial resolution of 0.5 and 2.0 m for panchromatic and multispectral images respectively. Images were checked and geo-referenced with the LiDAR data.

Ecological Vegetation Classes (EVCs), which describe the spatial extent of native vegetation species, were introduced by the Victorian Department of Sustainability and Environment in the 1990s. EVC mapping was implemented as part of the regional forest agreements, driven by a need to determine boundaries for a forest reserve system. The EVC mapping was undertaken first by the interpretation of aerial photographs and the process was designed to outline native vegetation patches and any obviously related patterns. The range of aerial photograph patterns was then field checked and lists of plant species recorded (Davies et al. 2002; Boyle and Lowe 2004). EVCs are the basic regional-scale mapping unit used for forest ecosystem assessments, biodiversity planning and conservation management in Victoria. For this study, the EVC data (which include data on plantation forests) provided by the HVP Plantations Pty Ltd were used as ground reference data.

3 Methods

3.1 *Canopy Height Modelling and LiDAR Data Normalization*

A digital elevation model (DEM) with two-metre horizontal resolution (grid size) was generated using the LiDAR ground data. Only the highest points of the first returns of non-ground LiDAR data within grid cells which represent the laser returns from the top of tree canopy were used to generate a canopy surface model (CSM) for the study area. A canopy height model (CHM) was then computed by subtracting the DEM from the CSM. The CHM representing the height variance of the top canopy will be used as data layer in the OBIA. However, the canopy top data alone cannot provide sufficient information to depict the multi-layered canopy structure. The use of LiDAR point clouds distributed throughout the whole canopy structure has been recognised as an effective way for quantifying the forest canopy structure. The height of laser returns above the ground is calculated from the difference between the laser returns (including those from tree canopy and understorey vegetation) and the corresponding DEM value. It is these normalised laser returns that provide more sufficient information to depict the vertical structure of the forests than the CHM does. The vertical distribution and configuration of the forest components are forest type dependent (Brandtberg et al. 2003). The structural differences will affect the distribution of the laser returns from the forests (Ørka et al. 2009; Antonarakis et al. 2008). Therefore, the metrics or variables derived from the LiDAR data can be used for tree species identification and forest type classification.

3.2 *Variables Derived from the Height Distribution of LiDAR Points*

In order to quantitatively describe the height distribution of LiDAR data in support of the forest classification, a grid of square columns with 4×4 m horizontal resolution (grid size) covering the study area were generated using Python programming language in ArcGIS 10.0 software. The none-ground LiDAR points within each of these square columns were extracted to quantify the vertical distribution of LiDAR points by calculating the variables of mean, standard deviation, coefficient of variation (CV), skewness, and kurtosis. The mean is a measure of the average height of LiDAR points within each of square columns and can be calculated by Eq. (1).

$$\bar{h} = \frac{\sum_{i=1}^{n} h_i}{n} \quad (1)$$

where h_i is normalized height value of a LiDAR points, and n is the number of LiDAR points in a square column.

Standard deviation, calculated using Eq. (2), measures the height variation of LiDAR points around the mean height within a square column, indicating how spread out a distribution is.

$$s = \sqrt{\frac{\sum_{i=1}^{n} (h_i - \bar{h})^2}{n-1}} \quad (2)$$

For two or more distributions with different mean heights, however, the relative height spread may not be determined by only comparing standard deviation (Groebner et al. 2011). Therefore, the CV is used to measure the relative variation of height distribution of LiDAR points within each square column. The CV is calculated as the ratio of the standard deviation to the mean using Eq. (3), expressed as a percentage.

$$CV = \frac{s}{\bar{h}} \times 100\,\% \quad (3)$$

Skewness is a measure of the asymmetry (compared to a normal distribution) of the height distribution of LiDAR points within a square column, and is calculated using Eq. (4) (Walford 2011).

$$Skewness = \frac{\sum_{i=1}^{n} (h_i - \bar{h})^3}{(n-1)s^3} \quad (4)$$

Kurtosis is a measure of peakedness relative to a normal distribution (Seltman 2010) and can be calculated by Eq. (5) (Walford 2011).

$$Kurtosis = \frac{\sum_{i=1}^{n} (h_i - \bar{h})^4}{ns^4} \quad (5)$$

3.3 Preparation of Input Data Layers

All these LiDAR-derived variables provide measures of forest canopy structure and density. The values of each of these variables over the study area are then represented with a raster image with 4 m resolution. The five raster 'images' created from the mean, standard deviation, CV, skewness, and kurtosis of the LiDAR height distribution, together with CHM and WorldView-2 multispectral bands were used as input data layers in the OBIA. The data layers and description are listed in Table 1.

3.4 Segmentation

The eCognition Developer 8.7 software was used for image segmentation. The segmentation algorithm implemented in the eCognition software is an optimization procedure which consecutively merges pixels or existing image objects. It is a bottom up segmentation approach, based on a pairwise region merging technique, that minimizes the average heterogeneity and maximizes segment homogeneity (Trimble eCognition 2011). The segmentation partitions an image into homogeneous multi-pixel regions (objects) based on several user-defined parameters. The output of the segmentation process can be adjusted through specification and weighting of input data and definition of parameters affecting the size, spectral homogeneity, spatial homogeneity, and shape of the resulting image objects (Chubey et al. 2006; Trimble eCognition 2011). In eCognition, the scale parameter, colour/shape ratio, smoothness/compactness ratio, and weights of input layers must be set to create meaningful objects (Trimble eCognition 2011). Five image segmentation schemes using different input data layers were implemented:

Table 1 Input data layers used in segmentations

Data sources	Data layers	Description
LiDAR	CHM	LiDAR-derived canopy height model
	Mean	Mean height of LiDAR points within each of square columns
	S	Standard deviation of LiDAR point heights within each of square columns
	CV	Coefficient of variation of LiDAR point heights within each of square columns
	Skew	Skewness of the height distribution of LiDAR points within each of square columns
	Kurt	Kurtosis of the height distribution of LiDAR points within each of square columns
WorldView-2	B	Blue band
	G	Green band
	R	Red band
	NIR1	Near infra-red band 1
	CB	Coastal blue band
	Y	Yellow band
	RE	Read-edge band
	NIR2	Near infra-red band 2

- Segmentation using LiDAR-derived data layers only;
- Segmentation using four conventional multispectral bands only (blue, green, red, and NIR);
- Segmentation using all eight multispectral bands including four new World-View-2 multispectral bands (coastal blue, yellow, red-edge and NIR2);
- Segmentation using both LiDAR-derived data layers and four conventional multispectral bands;
- Segmentation using both LiDAR-derived data layers and eight multispectral bands.

The scale parameter is an abstract term which determines the maximum allowed heterogeneity within an object (Trimble eCognition 2011). Larger scale parameters produce larger objects, and smaller scale parameters produce smaller objects (Benz et al. 2004). In recognition that the scale is one of the most critical factors that affect the segmentation results, there have been a number of attempts to determine optimal segmentation (Wang 2004; Kim et al. 2008; Radoux and Defourny 2008; Hay et al. 2001; Möller et al. 2007; Clinton et al. 2010; Kim et al. 2009). These methods may be applicable to the accuracy assessment of specific object detection (e.g., buildings, trees), but may not be suitable for accuracy assessment of thematic land cover maps from OBIA results (Radoux et al. 2011). So far, there are no specific guidelines on this issue. Proposed methods for optimal scale selection are not yet fully implemented in mainstream OBIA software, e.g., eCognition (Duro et al. 2012). Consequently, the most commonly used approach to the selection of scale parameters is still based on visual inspection of resultant objects through trial-and-error (Navulur 2007; Ke et al. 2010; Kim et al. 2008; Hay et al. 2005). In this study, a range of scale parameters were tested and visually evaluated for each of the segmentation schemes through the trial-and-error approach.

3.5 Object-Based Classification and Accuracy Assessment

Object-based variables were derived from all data layers including six LiDAR-derived layers and eight spectral bands for the classifications. In addition to the mean and standard deviation of each object in a layer, grey level co-occurrence matrix (GLCM) textures (homogeneity, contrast, dissimilarity, and correlation) were also calculated in the eCognition for all data layers. Details about GLCM textures can be seen in Haralick (1986).

It has been suggested that segmented objects should be used as sampling units for classifier training and accuracy assessment (Radoux et al. 2011) because the objects, instead of pixels are the basic units in object-based classification. Reference points over the EVC map were randomly selected in ArcGIS software. These reference points were then imported into eCognition software as a thematic layer and overlaid on the segmented objects in support of object sampling. Any object that contains reference points and overlaps with an EVC map polygon by more than 80 % of the object's own area was selected as a sample object. Using the same set of reference points for objects sampling provides similar sampling locations to minimise the impacts of different segments from five object segmentation schemes on the comparisons of classification results. Sample objects were divided to training objects and test objects in object-based classifications. The training objects were used to train the classifier while the test objects were used to assess the classification accuracy.

Object-based classification was implemented with decision trees in CART 6.0 software (Steinberg and Golovnya 2006). Decision trees were developed by Breiman et al. (1984) and have been quickly recognised as a valuable tool for discriminating complex relationships among natural environmental variables (Kagan et al. 2006). A decision tree is constructed with a recursive partitioning algorithm from a training data set. In the case of binary trees such as in CART, the training set is split into two subsets in each tree node (Steinberg and Golovnya 2006). The classification process is implemented by a set of *if-then* rules that determine the path to be followed and recursively split the data until no further splits are possible (Lawrence and Wright 2001; Tso and Mather 2009). In order to avoid over-fitting the decision tree model, a pruning procedure was automatically implemented in CART. The pruning process trims the tree by removing the splits and branches that are least useful to produce an optimal decision tree (Steinberg and Golovnya 2006). Five decision trees that are corresponding to the five segmentation schemes were constructed and implemented for object-based forest classification.

The overall accuracy, Kappa values, producer's accuracy and user's accuracy were calculated from confusion matrix (Congalton and Green 2009) to assess the accuracy of each of the classification results. The z-statistic was calculated using Eq. (6) to test whether the Kappa values from two classifications are significantly different from each other to evaluate the contribution of different data layers to the classifications:

$$Z = \frac{k_1 - k_2}{\sqrt{Var(k_1) + Var(k_2)}} \tag{6}$$

where k_1 and k_2 are Kappa values and $Var(k_1)$ and $Var(k_2)$ are their variances calculated according to Congalton and Green (2009). The Z is a standard normal distribution variable. The hypotheses to be tested are:

H_0 $k_1 = k_2$ (the two classification results are not significantly different)

H_1 $k_1 \neq k_2$ (the two classification results are significantly different)

H_0 is rejected if $Z > |Z_{\alpha/2}|$, where $\alpha/2$ is the confident level of the two-tailed Z test (Congalton and Green 2009).

4 Results and Discussion

4.1 Segmentation and Scale Parameter

The segmentation results from different input data including LiDAR data only, four WorldView-2 bands only, eight WorldView-2 bands, LiDAR data and four WorldView-2 bands, and LiDAR data and eight WorldView-2 bands were obtained using selected scale parameters through tests and visual evaluation. The scale parameters selected for different segmentation schemes were listed in Table 2, ranging from 60 to 70. The values of 0.9 and 0.1 were chosen for colour and shape criteria while the values of 0.7 and 0.3 are used for compactness and smoothness criteria in eCognition, and all data layers were assigned equal weight in all five segmentation schemes. The number of segmented objects varied with the different input data and scale parameter, but using the above parameter values, small forest patches can be delineated with the segmented objects. The scale parameters and the number of segmented objects using different input data layers were listed in Table 2.

The estimation of an optimal segmentation scale is a critical but challenging issue in the use of OBIA. The attempts to use predefined object-specific spatial measures to automate, and guide the selection of the scale parameter have been largely unsuccessful because there is no recognizable relationship between the scale parameter in currently available OBIA software such as eCognition and spatial measures specific to the image-objects composing a scene (Hay et al. 2005). Castilla and Hay (2008) described a two-sided problem in image segmentation: over-segmentation referring to a situation where the contrast between some adjacent segments is insufficient and should be merged into a single image-object, and under-segmentation referring to the existence of segments that lack coherency and should be split into smaller objects. In general, over-segmentation is less serious than under-segmentation because the over-segmentation problem can be corrected in subsequent classification stage by aggregating over-segmented objects (Castilla and Hay 2008; Radoux et al. 2011), so a good segmentation is one that shows little over-segmentation and no under-segmentation (Castilla and Hay 2008). In practice, in particular for land cover classification, the image segmentation process is considered complete once image objects visually correspond to meaningful real world objects of interest (Duro et al. 2012; Radoux et al. 2011).

Table 2 The scale parameters and the number of segmented objects using different input data

Data source	LiDAR data	4 image bands	8 image bands	LiDAR and 4 image bands	LiDAR and 8 image bands
Scale parameter	60	70	65	70	60
Object numbers	845	1,066	846	911	789

4.2 Accuracy Assessment of Classifications

Classification accuracy was assessed using test sampling objects. The overall accuracy and Kappa values from five classifications were listed in Table 3. It is seen in Table 3 that an overall classification accuracy of 61.39 % was achieved using only the LiDAR data. The overall accuracy obtained from the classification using four conventional WorldView-2 bands was nearly same to the accuracy from using LiDAR data. The use of eight image bands increased the overall accuracy to 70.40 %, implying the contribution of four new WorldView-2 bands to the classification. With the inclusion of LiDAR data to spectral bands, the overall classification accuracy was improved. For example, the overall accuracy increased from 61.42 % (using only four image bands) to 73.50 %. The Kappa values in Table 3 show the similar trend of classification accuracy when using different data sets. The combined use of the LiDAR data and eight image bands produced the highest overall accuracy. This indicated that the integration of the LiDAR data with eight WorldView-2 bands has more discriminatory power in object-based forest classification.

Table 4 summaries the Kappa *Z*-test statistics of the classifications, showing whether the accuracy of one classification was significantly different from another. If the *Z*-value in Table 4 is greater than 1.96 (the *Z*-critical value at 95 % confidence level), the two classifications are considered to be significantly different from a statistical point of view. The classification based either on eight image bands or on the LiDAR data and four image bands did not show a statistically significant difference with the classification using the LiDAR data only although the differences can be observed in Table 4 in terms of the overall accuracy and the Kappa values. The inclusion of the LiDAR data into either four image bands or eight image bands did significantly improve the classification accuracy compared with the classifications using four image bands or eight image bands only. It was evident that the combination of the LiDAR data and eight WorldView-2 image

Table 3 Overall accuracy and Kappa values from five classifications using different input data

Data source	LiDAR	4 image bands	8 image bands	LiDAR and 4 image bands	LiDAR and 8 image bands
Overall accuracy (%)	61.39	61.42	70.40	73.50	82.35
Kappa	0.51	0.49	0.62	0.66	0.77

Table 4 Kappa *Z*-test statistics of the classifications. *Z*-values in bold indicate significant difference between the two classifications

Data source	LiDAR	4 image bands	8 image bands	LiDAR and 4 image bands
4 image bands	0.21			
8 image bands	1.38	1.70		
LiDAR and 4 image bands	1.92	**2.27**	0.59	
LiDAR and 8 image bands	**3.50**	**3.99**	**2.30**	1.67

bands produced significantly higher classification accuracy than the LiDAR data or the four multispectral bands did. The forest species classification map from using both LiDAR and eight WorldView-2 multispectral bands are shown in Fig. 2.

Figure 3 shows the producer's and user's accuracies of the classifications using different data sets, representing accuracies for individual forest types (Congalton and Green 2009). Producer's accuracy indicates how well test sampling objects of a given forest type are classified (Lillesand et al. 2008; Jensen 2005). Figure 3a shows t hat four image bands or eight image bands alone cannot achieve a producer's accuracy better than 50 % for mixed forest and Silver Wattle. The producer's accuracies resulted from the LiDAR data for mixed forest and Silver Wattle were higher than those from image bands. Four image bands, LiDAR data, or even the combination of these two data sets cannot create a higher producer's accuracy for Mountain Ash. However, eight image bands alone did produce a higher producer's accuracy, indicating the strong capability of new WorldView-2 bands for capturing spectral characteristics of Mountain Ash. The WorldView-2 data (four bands or eight bands) also showed strong capability of identifying the rainforest. Integration of the LiDAR data and eight image bands produced higher producer's accuracies for all forest types. It is seen that a high producer's accuracy

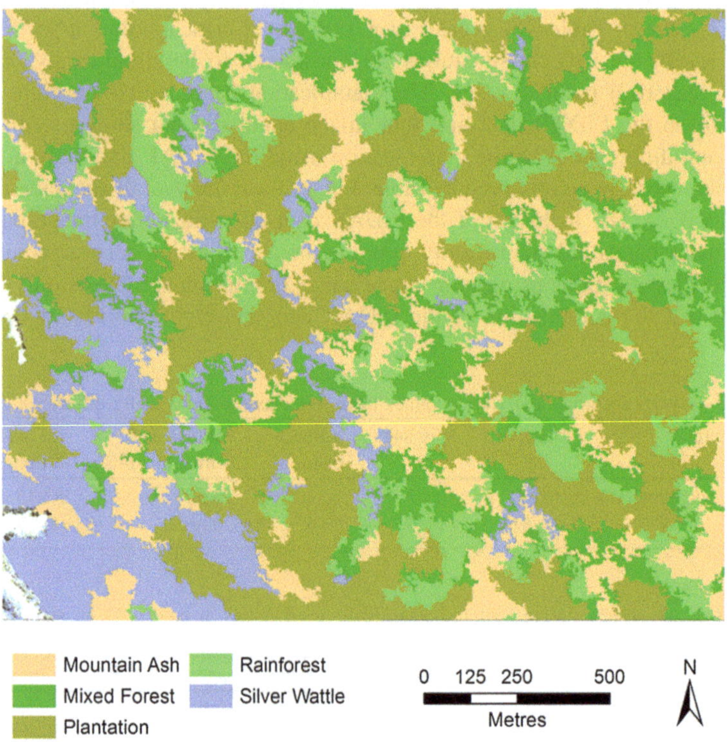

Fig. 2 Forest species classification map from using both LiDAR and eight WorldView-2 *multispectral bands*

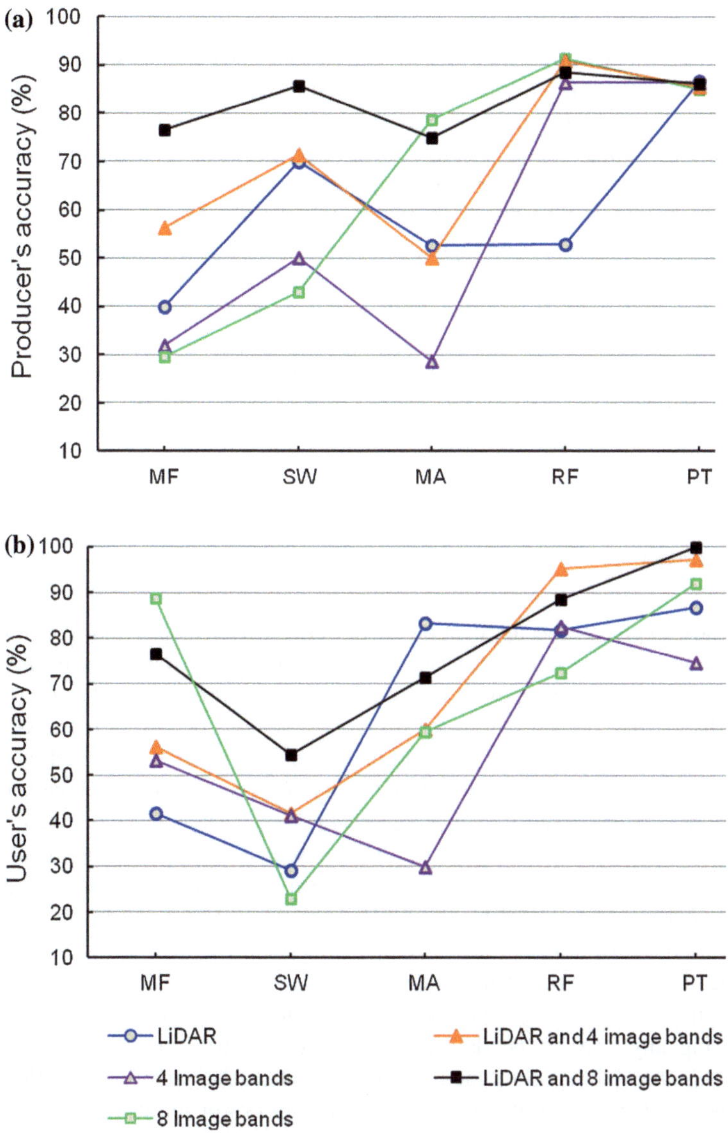

Fig. 3 Producer's and user's accuracies of the classifications using different input data for mixed forest (*MF*), silver wattle (*SW*), mountain ash (*MA*), rainforest (*RF*), and plantation (*PT*): **a** producer's accuracy, and **b** user's accuracy

for plantation forests can be achieved using any of the data sets, implying that the characteristics of spectral reflection and vertical structure of plantation forests can be discriminated from other forests by LiDAR and/or WorldView-2 image data.

Figure 3 indicates the probability that a test sampling object classified into a given forest type actually represents that forest type on the ground (Lillesand et al.

2008; Jensen 2005). Similar to producer's accuracies, user's accuracies varied with forest types as shown in Fig. 3b. The highest user's accuracies were achieved for rainforest and plantation forests by integration of the LiDAR data with either four image bands or eight image bands. Combined use of the LiDAR data and WorldView-2 image data produced both high producer's accuracies and high user's accuracies for rainforest and plantation forests. Therefore, the classified rainforest and plantation forest in this study are considered to be very reliable. Although Fig. 3a showed the higher producer's accuracies for Silver Wattle using the LiDAR and image data, relative low user's accuracies were observed. It indicated that over 85 % of test sampling objects of Silver Wattle were correctly classified. On the other hand, however, only 55 % of the objects that were classified into Silver Wattle were Silver Wattle. Considering all forest types, it is evident that the integration of the LiDAR data and eight WorldView-2 bands produced overall better user's accuracies.

5 Conclusions

This study explored the application of airborne LiDAR data and WorldView-2 satellite imagery data in object-based forest species classifications in an Australian cool temperate rainforest environment. Using normalised LiDAR point clouds above the ground, the vertical forest structure was described using LiDAR-derived variables based on the height distribution of LiDAR points within each of square columns. These variables were subsequently used as input data layers in segmentations. Object-based variables generated from these data layers were ranked to be important by decision trees in object-based forest species classifications in our study area. New WorldView-2 multispectral bands, in particular the red-edge, yellow, and near infrared-2 bands did show their contributions to the classifications. Based on overall accuracies, Kappa values and Kappa Z-test statistics of the classifications, it is evident that integration of the LiDAR data and eight WorldView-2 multispectral bands produced significantly higher classification accuracy than using any one of the data sets alone. In addition, producer's and user's accuracies showed that combined use of the LiDAR data and eight WorldView-2 multispectral bands exhibited more discriminatory power for all five forest species in the study area, including cool temperate rainforest dominated by Myrtle Beech (*N. cunninghamii*), Mountain Ash (*E. regnans*), mixed forest consisting of overstorey Mountain Ash and understorey Myrtle Beech, Silver Wattle (*A. dealbata*), and plantation dominated by Shining Gum (*E. nitens*). Success of cool temperate rainforest and adjacent forest species classification by the methods demonstrated in this study strongly warrants the adoption of airborne LiDAR data and high spatial resolution satellite imagery in the management of Australia's forestry resources.

References

Adam P (1992) Australian rainforests. Oxford University Press, Oxford

Antonarakis AS, Richards KS, Brasington J (2008) Object-based land cover classification using airborne LiDAR. Remote Sens Environ 112:2988–2998

Arroyo LA, Johansen K, Armston J, Phinn S (2010) Integration of LiDAR and QuickBird imagery for mapping riparian biophysical parameters and land cover types in Australian tropical savannas. For Ecol Manage 259:598–606

Baatz M, Schäpe A (2000) Multi-resolution segmentation: an optimization approach for high quality multi-scale image segmentation. In: Strobl J, Blaschke T (eds) Angewandte geographische informationsverarbeitung. Wichmann-Verlag, Heidelberg, pp 12–23

Benz UC, Hofmann P, Willhauck G, Lingenfelder I, Heynen M (2004) Multi-resolution, object-oriented fuzzy analysis of remote sensing data for GIS-ready information. ISPRS J Photogram Remote Sens 58:239–258

Blaschke T (2010) Object based image analysis for remote sensing. ISPRS J Photogram Remote Sens 65:2–16

Borak JS, Strahler AH (1999) Feature selection and land cover classification of a MODIS-like data set for a semiarid environment. Int J Remote Sens 20(5):919–938

Boyle C, Lowe KW (2004) Biodiversity action planning strategic overview for the Strzelecki Ranges bioregion. Victorian Department of Sustainability and Environment, Melbourne

Brandtberg T, Warner TA, Landenberger RE, McGraw JB (2003) Detection and analysis of individual leaf-off tree crowns in small footprint, high sampling density lidar data from the eastern deciduous forest in North America. Remote Sens Environ 85:290–303

Breiman L, Friedman JH, Olsen RA, Stone CJ (1984) Classification and regression trees. Wadsworth International Group, Belmont

Brown de Colstoun ECB, Story MH, Thompson C, Commisso K, Smith TG, Irons JR (2003) National park vegetation mapping using multi-temporal landsat 7 data and a decision tree classifier. Remote Sens Environ 85:316–327

Castilla G, Hay GJ (2008) Image objects and geographic objects. In: Blaschke T, Lang S, Hay GJ (eds) Object-based image analysis. Springer, Berlin, pp 91–110

Chubey MS, Franklin SE, Wulder MA (2006) Object-based analysis of Ikonos-2 imagery for extraction of forest inventory parameters. Photogram Eng Remote Sens 72(4):383–394

Clinton N, Holt A, Scarborough J, Yan L, Gong P (2010) Accuracy assessment measures of object-based image segmentation goodness. Photogram Eng Remote Sens 76(3):289–299

Congalton RG, Green K (2009) Assessing the accuracy of remotely sensed data: principles and practices, 2nd edn. CRC Press, Boca Raton

Danson FM, Morsdorf F, Koetz B (2009) Airborne and terrestrial laser scanning for measuring vegetation canopy structure. In: Heritage GL, Large ARG (eds) Laser scanning for the environmental sciences. Wiley-Blackwell, West Sussex, pp 201–219

Davies JB, Oates AM, Trumbull-Ward AV (2002) Ecological vegetation class mapping at 1:25000 in Gippsland. Victorian Department of Natural Resources and Environment, Melbourne

DigitalGlobe (2009) White Paper—the benefits of the 8 spectral bands of WorldView-2. DigitalGlobe, Inc, Longmont, CO

DSE (2011) EVC benchmarks—Strzelecki Ranges bioregion. Victorian Department of Sustainability and Environment, Melbourne

Duro DC, Franklin SE, Dubé MG (2012) A comparison of pixel-based and object-based image analysis with selected machine learning algorithms for the classification of agricultural landscapes using SPOT-5 HRG imagery. Remote Sens Environ 118:259–272

Erdody TL, Moskal LM (2010) Fusion of LiDAR and imagery for estimating forest canopy fuels. Remote Sens Environ 114:725–737

Evans DL, Roberts SD, Parker RC (2006) LiDAR—A new tool for forest measurements? Forest Chronicle 82(2):211–218

Franklin SE (2001) Remote sensing for sustainable forest management. CRC Press LLC, Boca Raton, London, New York and Washington, DC

Friedl MA, Brodley CE (1997) Decision tree classification of land cover from remotely sensed data. Remote Sens Environ 61:399–409

Groebner DF, Shannon PW, Fry PC, Smith KD (2011) Business statistics: a decision-making approach, 8th edn. Pearson Education, Inc., Boston

Hansen M, Dubayah R, Defries R (1996) Classification trees: an alternative to traditional land cover classifiers. Int J Remote Sens 17(5):1075–1081

Haralick RM (1986) Statistical image texture analysis. In: Young TY, Fu K-S (eds) Handbook of pattern recognition and image processing. Academic, Inc., San Diego, pp 247–279

Hay GJ, Castilla G, Wulder MA, Ruiz JR (2005) An automated object-based approach for the multiscale image segmentation of forest scenes. Int J Appl Earth Obs Geoinf 7:339–359

Hay GJ, Marceau DJ, Bouchard A, Dubá P (2001) A multiscale framework for landscape analysis: object-specific upscaling. Landscape Ecol 16:471–490

Hay GJ, Niemann KO, McLean GF (1995) An object-specific image-texture analysis of H-resolution forest imagery. Remote Sens Environ 55:108–122

Holmgren J, Persson Å, Söderman U (2008) Species identification of individual trees by combining high resolution LiDAR data with multi-spectral images. Int J Remote Sens 29(5):1537–1552

Jensen JR (2005) Introductory digital image processing: a remote sensing perspective, 3rd edn. Prentice Hall, New Jersey

Kagan JS, Ohmann JA, Gregory MJ, Tobalske C, Hak JC, Fried J (2006) Final report on land cover mapping methods: map zones 8 and 9, Pacific Northwest ReGAP. Institute for Natural Resources, Oregon State University, Corvallis, OR

Ke Y, Quackenbush LJ, Im J (2010) Synergistic use of QuickBird multispectral imagery and Lidar data for object-based forest species classification. Remote Sens Environ 114:1141–1151

Kim M, Madden M, Warner T (2008) Estimation of optimal image object size for the segmentation of forest stands with multispectral IKONOS imagery. In: Blaschke T, Lang S, Hay GJ (eds) Object-based image analysis. Springer, Berlin, pp 291–307

Kim M, Madden M, Warner TA (2009) Forest type mapping using object-specific texture measures from multispectral Ikonos imagery: segmentation quality and image classification issues. Photogram Eng Remote Sens 75(7):819–829

Lawrence R, Bunn A, Powell S, Zambon M (2004) Classification of remotely sensed imagery using stochastic gradient boosting as a refinement of classification tree analysis. Remote Sens Environ 90:331–336

Lawrence RL, Wright A (2001) Rule-based classification systems using classification and regression tree (CART) analysis. Photogram Eng Remote Sens 67(10):1137–1142

Legg S (1986) Farm abandonment in South Gippsland's Strzelecki Ranges, 1870–1925: challenge or tragedy. Gippsland Heritage J 1(1):14–22

Lillesand TM, Kiefer RW, Chipman JW (2008) Remote sensing and image interpretation, 6th edn. Wiley, New York

Lim K, Treitz P, Wulder M, St-Onge B, Flood M (2003) LiDAR remote sensing of forest structure. Prog Phys Geogr 27(1):88–106

Lovell LL, Jupp DLB, Culvenor DS, Coops NC (2003) Using airborne and ground-based ranging lidar to measure canopy structure in Australian forests. Can J Remote Sens 29(5):607–622

Mallinis G, Koutsias N, Tsakiri-Strati M, Karteris M (2008) Object-based classification using QuickBird imagery for delineating forest vegetation polygons in a Mediterranean test site. ISPRS J Photogram Remote Sens 63:237–250

Michaelsen J, Schimel DS, Friedl MA, Davis FW, Dubayah RC (1994) Regression tree analysis of satellite and terrain data to guide vegetation sampling and surveys. J Veg Sci 5:673–686

Möller M, Lymburner L, Volk M (2007) The comparison index: a tool for assessing the accuracy of image segmentation. Int J Appl Earth Obs Geoinf 9:311–321

Morsdorf F, Mårell A, Koetz B, Cassagne N, Pimont F, Rigolot E, Allgöwer B (2010) Discrimination of vegetation strata in a multi-layered Mediterranean forest ecosystem using

height and intensity information derived from airborne laser scanning. Remote Sens Environ 114:1403–1415
Navulur K (2007) Multispectral image analysis using the object-oriented paradigm. CRC Press, Boca Raton
Noble WS (1978) The Strzeleckis: a new future for the heartbreak hills. Victoria Forests Commission, Melbourne
Novichikhin A (2011) Developent of object-oriented image classification technique with a tree heights and species interpretation using 8-band VHR satellite imagery. Faculty of Geography, Lomonosov Moscow State University, Moscow
Ørka HO, Næsset E, Bollandsås OM (2009) Classifying species of individual trees by intensity and structure features derived from airborne laser scanner data. Remote Sens Environ 113:1163–1174
Ozdemir I, Karnieli A (2011) Precicting forest structural parameters using the image texture derived from WorldView-2 multispectral imagery in a dryland forest, Israel. Int J Appl Earth Obs Geoinf 13:701–710
Pal M, Mather PM (2003) An assessment of the effectiveness of decision tree methods for land cover classification. Remote Sens Environ 86:554–565
Peel B (1999) Rainforests and Cool Temperate Mixed Forests of Victoria. Department of Natural Resources and Environment, East Melbourne
Pekkarinen A (2002) Image segment-based spectral features in the estimation of timber volume. Remote Sens Environ 82:349–359
Popescu SC, Wynne RH, Nelson RF (2002) Estimating plot-level tree heights with lidar: local filtering with a canopy-height based variable window size. Comput Electron Agric 37:71–95
Potapov P, Hansen MC, Stehman SV, Loveland TR, Pittman K (2008) Combining MODIS and landsat imagery to estimate and map boreal forest cover loss. Remote Sens Environ 112:3708–3719
Radoux J, Bogaert P, Fasbender D, Defourny P (2011) Thematic accuracy assessment of geographic object-based image classification. Int J Geog Inform Sci 25(6):895–911
Radoux J, Defourny P (2008) Quality assessment of segmentation results devoted to object-based classification. In: Blaschke T, Lang S, Hay GJ (eds) Object-based image analysis. Springer, Berlin, pp 257–271
Riaño D, Meier E, Allgöwer B, Chuvieco E, Ustin SL (2003) Modeling airborne laser scanning data for the spatial generation of critical forest parameters in fire behavior modeling. Remote Sens Environ 86:177–186
Sasaki T, Imanishi J, Ioki K, Morimoto Y, Kitada K (2010) Object-based classification of land cover and tree species by integrating airborne LiDAR and high spatial resolution imagery data. Landscape Ecol Eng. doi:10.1007/s11355-011-0158-z
Scanlan I, McElhinny C, Turner P (2010) A methodology for modelling canopy structure: an exploratory analysis in the tall wet eucalypt forests of southern Tasmania. Forests 1:4–24
Seltman HJ (2010) Experimental design and analysis. Carnegie Mellon University, Pittsburgh
St-Onge BA, Cavayas F (1997) Automated forest structure mapping from high resolution imagery based on directional semivariogram estimates. Remote Sens Environ 61:82–95
Steinberg D, Golovnya M (2006) CART 6.0 user's manual. Salford Systems, San Diego, CA
Trimble eCognition (2011) eCognition developer 8.7 user guide. Trimble Germany GmbH, Munich
Tso B, Mather PM (2009) Classification methods for remotely sensed data, 2nd edn. CRC Press, Boca Raton
Walford N (2011) Practical statistics for geographers and earth scientists. Wiley-Blackwell, Oxford
Wang L (2004) Integration of object-based and pixel-based classification for mapping mangroves with IKONOS imagery. Int J Remote Sens 25(24):5655–5668
WGCMA (2009) Strzelecki Ranges bioregion Grand Ridge biodiversity landscape plan. West Gippsland Catchment Management Authority, Traralgon

Yu Q, Gong P, Clinton N, Biging G, Kelly M, Schirokauer D (2006) Object-based detailed vegetation classification with airborne high spatial resolution remote sensing imagery. Photogram Eng Remote Sens 72(7):799–811

Zhang Z, Liu X (2012) Support vector machines for tree species identification using LiDAR-derived structure and intensity variables. Geocarto Int. doi:10.1080/10106049.2012.710653

Zhang Z, Liu X, Peterson J, Wright W (2011) Cool temperate rainforest and adjacent forests classification using airborne LiDAR data. Area 43(4):438–448

Unmanned Aerial Vehicle Photogrammetric Results Using Different Real Time Kinematic Global Positioning System Approaches

Khairul Nizam Tahar, Anuar Ahmad, Wan Abdul Aziz Wan Mohd Akib and Wan Mohd Naim Wan Mohd

Abstract This chapter describes photogrammetric results based on unmanned aerial vehicle (UAV) images for slope mapping. This chapter discussed the effect of photogrammetric results based on different observation time and network adjustment of Real Time Kinematic Global Positioning System (RTK GPS) in photogrammetric image processing. A complete set of fixed wing UAV with autonomous flight was used to capture image from certain altitude at the study area. All acquired images were processed using photogrammetric software. Two primary results were produced in this study namely digital elevation model and digital orthophoto. Analyses were carried out to determine the best RTK GPS data based on time and network during photogrammetric image processing. It was found that, RTK GPS adjustment from known point with 10 min observation time gives greater accuracy compared to the local network adjustment. In conclusion, UAV has potential in the slope mapping because it can provide high accuracy photogrammetric results.

Keywords Slope mapping · UAV · Photogrammetry · RTK GPS

1 Introduction

In recent years, the Terrestrial Laser Scanning (TLS) and Airborne Laser Scanning (ALS) have acquired a rapidly growing interest for three dimensional data acquisition in the remote sensing and photogrammetry field. These laser

K. N. Tahar (✉)
Department of Geomatic Science and Surveying, Universiti Teknologi MARA, 40450 MARA, Shah Alam, Malaysia
e-mail: nizamtahar@gmail.com

A. Ahmad · W. A. A. W. M. Akib · W. M. N. W. Mohd
Department of Geoinformation, Faculty of Geoinformation and Real Estate, Universiti Teknologi Malaysia, 81310 Johor Bahru, Johor, Malaysia

A. Abdul Rahman et al. (eds.), *Developments in Multidimensional Spatial Data Models*, Lecture Notes in Geoinformation and Cartography,
DOI: 10.1007/978-3-642-36379-5_8,

techniques measure distances from the platform derives from the time by the laser beam. In many studies, laser scanner was proven to provide high accurate result in land deformation (Tarchi et al. 2003; Teza et al. 2007; Abellan et al. 2010; Yue et al. 2010; Gi al. 2000; Moss 2000; Pesci et al. 2005; Squarzoni et al. 2005; Al-Bayari and Saleh 2007; Li et al. 2010; Cardenal et al. 2006; Pocsai et al. 2011; Teza et al. 2008. In theory, TLS and ALS provide a very high data redundancy and it is suitable for slope monitoring in centimeter level. However, TLS and ALS platform involves huge cost to complete the mission and these methods are only suitable for large area and not practical for small project with low budget. There are also many previous techniques for data acquisition in surveying field such as close range photogrammetry, real time kinematic global positioning system, satellite image, leveling, and total station. All of these techniques offer a different results in terms of accuracy, resolution and cost.

Terrestrial laser scanning (TLS) is a ground-based technique for collecting high-density 3D geospatial data. It is an active imaging system whereby laser pulses are emitted by the scanner and observables include the range and intensity of pulse returns reflected by the surface or object being scanned. TLS measurements, combined with the orientation and position of the scanner, produce a 3-dimensional "point cloud" dataset. The primary capability of TLS is the generation of high resolution 3D maps and images of surfaces and objects over scales of meters to kilometers with centimeter to sub-centimeter precision. It is a powerful geodetic imaging tool ideal for supporting a wide spectrum of user applications in many different environments (Johan et al. 2010). The incorporation of GPS measurements provides accurate georeferencing of TLS data in an absolute reference frame. The addition of digital photography yields photorealistic 3D images. The main advantage of TLS and ALS over other geodetic techniques is its property of direct, rapid and detailed capture of object geometry in 3D.

ALS using airplanes or helicopters constitutes one well established method for landslide surveying (Barber et al. 2001). The ALS measures the height of the ground surface and other features in large areas of landscape with a resolution and accuracy hitherto unavailable. ALS operates by using a pulsed laser beam which scans from side to side as the aircraft/helicopter flies over the area to be surveyed, measuring between 20,000 and 100,000 points per second to build an accurate, high resolution 3D-model of the ground and the features upon it. The ALS scanning device is mounted at the bottom or sides of an aircraft flying along selected parallel flight paths. The instrument is bundled with an onboard differential GPS and an inertial navigation system to locate accurately the position of the aircraft. The information, together with the orientation of the scanning mirror for firing the laser, can compute the geodetic reference of the objects scanned.

Aerial photogrammetry using manned aircraft has been explored many years ago and it is very efficient for large study area. Manned aerial photogrammetry is also used to update new areas which need to be included in the existing map. Previously, manned aerial photogrammetry uses film as the raw images of the earth surface but now it has been converted into digital images. Film images used analytical stereoplotter in order to determine the correct image position same as flight mission. Digital aerial images used the same process using digital photogrammetric software. Manned aerial photogrammetry is widely used to classify land use at study area. Close range photogrammetry is one of the methods in mapping field. Close range photogrammetry is suitable for small area or focused at the specific object to fulfill the project needs. Close range photogrammetry is used in many applications such as cultural heritage recording and in architectural surveying, but they still have some deficiencies. First, image processing and orientation are needed before the model can be analyzed. Second, they are not suitable for environments with quite unfavorable external conditions (insufficient illumination, dust, etc.). It is thus necessary to apply another technique that allows direct 3D measurement of the object geometry with high accuracy. The traditional way in both photogrammetry methods permit the 3D models of the terrain to be carried out and, by means of Digital Terrain Model (DTM) comparisons, to realize multi-temporal studies. The massive introduction of modern digital photogrammetric workstations, with automatic matching procedures, allows for a rapid DTM production.

GPS has two measurements namely code measurement and carrier phase measurement. Code measurement has point positioning and differential GPS positioning with the accuracy of 5–10 m and below 1–3 m respectively. Point positioning requires only one receiver and differential GPS needs radio or communication link. Carrier phase measurement has static post-processed and Real Time Kinematic (RTK). Static post-processed can give accuracy of up to 5 mm and RTK can give accuracy of 1–2 cm but limited to 10 km range (Janssen et al. 2011). Carried phase measurement requires expensive equipments but static post-processed needs more observation time while RTK only needs a few minutes to obtain high accuracy data. RTK GPS is very useful in finding existing survey marks, obtaining a high accuracy of point positioning, obtaining real time adjustment based on local network or master station, and providing an accurate baseline for survey control. RTK GPS consists of dual frequency GPS receivers and it is utilized with a UHF radio link (Montane and Torres 2006). This study focuses on the best RTK technique and observation time for unmanned aerial vehicle image processing.

In the last few years, unmanned aerial vehicle (UAV) has received an increasing interest as one of the reliable methods for slope studies. Since the reality is three-dimensional (3D), it is a great advantage to conduct modeling in 3D environment. Today, UAV has made it possible to efficiently process and visualize data in 3D. It is especially important to acquire fast and accurate 3D geometric and visual information with minimum costs. UAV is one of the surveying methods conceived years ago that in a short time can supply DSM (Digital Surface Model) and good quality DTM as a result of elaborations with specific procedures. The main advantage of UAV over traditional surveying techniques is its property of direct, rapid and detailed image capture of study area. Other advantages are the dramatic reduction in costs and much faster project completion, possibility to survey remotely very complex, inaccessible and hazardous objects and areas, where the traditional techniques fail.

In addition, the cost of UAV is much lower compared to aircraft or terrestrial equipments. Micro UAV has potential in forest and agricultural applications. It is because UAVs are more flexible and are able to obtain data in any weather condition. The advantages of UAVs are; low in cost, flexible, high resolution images, able to fly under cloud, easy to launch and land, and very safe (Lei et al. 2009). The disadvantages of UAVs include payload limitation, small coverage for one image, increasing numbers of images that need to be processed, and large geometric distortion. Due to the current technology, most UAVs are installed with the autonomous pilot chip. The autonomous UAV features include attitude stabilization, velocity and position control, rotor speed control, waypoint guidance as well as automatic take-off and landing. Real-time flight data such as position, speed, attitude, distance, auto-pilot functionality, and several safety features, including status indicators of the serial data-link connection between ground station and aircraft, GPS solution validity, RC radio connection, voltage of power supply and RC receivers/servos, fuel level indicator, and rotor speed indicator (Hai et al. 2010). An operator is only needed for the launching and landing of the UAV in order to reduce the damages of UAV.

Most researchers have developed and designed UAV for research and have investigated the capabilities of UAV in solving their problem (i.e., slope mapping, urban mapping, natural hazard monitoring and etc). UAV can be used in many applications such as archeology documentation, documentary video, national security, rescue and mapping (Tahar and Ahmad 2012). The aim of this study is to determine the effect of UAV photogrammetric results based on different Real Time Kinematic Global Positioning System (RTK-GPS) data. A different RTK-GPS observation has been conducted to analyze the accuracy of photogrammetric results.

2 The Methodology

The study was conducted at the slope area in Skudai, Johor, Malaysia. Figure 1 shows the location of the study area and its condition. The study area is located at latitude 1° 33′ 39″ and longitude 103° 39′ 13″. The reason this study area is selected because there are new development which is not yet updated in existing digital globe. UAV has been explored to update new information at the study area.

UAV flight line has been designed by using Lentsika software. Lentsika software required information such as estimated coordinates of the study area, pixel size of digital image, image resolution and estimated ground resolution (Tahar et al. 2011). The estimated coordinates at the study area were obtained from Google Earth coordinates. However, the interval distance between each images has

Fig. 1.1 Study area

been calculated based on user requirement. In this study, each pair of images overlapped about 60 % and side lap about 30 %. Ground coverage area for each image was calculated based on pixel size, image resolution and scale of aerial photograph. After that, flight line file was sent to horizon software for final inspection. Horizon software finalized the flight line at the study area and this software was also used to monitor UAV during flight mission. In this study, 102 exposure stations were designed to cover the study area in Skudai, Malaysia. The study area is located at the new development area in order to determine the capabilities of UAV in updating new information. Figure 2 shows an example of fixed wing UAV and ground control software during flight mission.

A fixed wing UAV, known as Cropcam is used to capture images at the study area. This UAV is very user friendly and is highly efficient for the commercial market. Cropcam is a radio control plane and it is equipped with a consumer digital camera (Pentax), autonomous navigation control board, and preprogrammed ground control software. Cropcam can fly autonomously or manually depending on the project needs. Cropcam main body includes eight foot wing, four foot length, electric engine with Axi brushless, camera mount, servos, propellers, and battery case. Cropcam UAV can fly up to 55 min with four thunder power L-polymer 2100 mah or equally to 11.1 V. Its average speed is 60 km/h and maximum wind 30 km/h. Cropcam UAV can be controlled by 2.4 Ghz radio

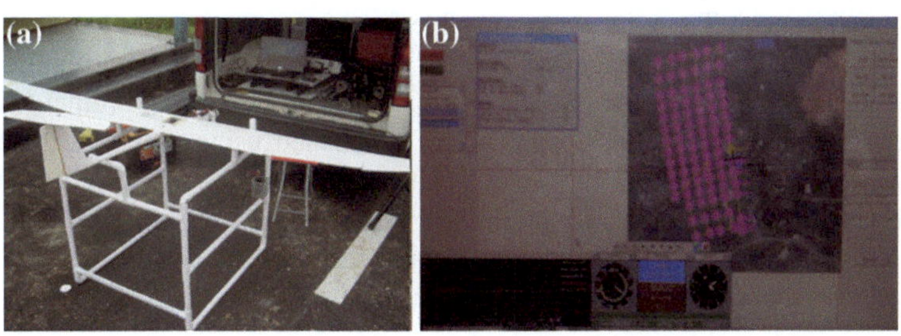

Fig. 1.2 UAV and software. **a** Fixed wing UAV. **b** Horizon software

modem within the range of 3 km or two miles. Cropcam flies up to 400–2,100 feet above the surface level and it depends on the mission requirement and approval from aviation department. An infrared L.E.D switch trigger is used to capture images automatically during flight mission. Professional operator is the most important person to ensure the safety of cropcam UAV during the launching and landing stages.

Before flight mission, an operator needs to test all UAV electronic parts to ensure that it is working and functioning well. This test is very important to avoid any problem on UAV during flight mission. An operator needs to test the electronic speed control, UAV body such as rudder, elevator and main frame, GPS onboard, camera mount, digital camera and propeller. After all tests have been done, an operator needs to find a suitable location for launching operation. In this study, we have applied autonomous flight mission to capture image at the study area. Basically, autonomous UAV need operator and ground crew station. In this study, an operator is responsible to control UAV during launching and landing operation to avoid any damages on UAV. Ground crew station is responsible to design the flight line at the study area and monitor UAV attitude, battery status, number of satellite, and data link between UAV and computer at the ground station. About 102 images were captured to cover the whole study area. The flight mission took about 15 min to finish.

All UAV images went through all the photogrammetric steps for example inner orientation, relative orientation, aerial triangulation and bundle adjustment. Inner orientation requires the input of camera parameters and it can be obtained from the camera calibration results. Camera calibration results can be carried out before or after the flight mission. There are several inputs that need to be included during camera calibration process such as number of horizontal and vertical pixel for each images, estimated focal length and pixel size. Pixel size can be calculated using the comparison between size of the image and the real size. Camera calibration results include all camera parameters such as focal length, principal distance for x and y, radial lens distortion, tangential lens distortion, and affinity. Relative orientation involves tie points measurements between overlapped pair of images or between strips. Tie points can be measured manually or automatically. Normally, automatic tie points give good and bad tie points and user needs to select the best tie points only to achieve good photogrammetric results. Automatic tie points measurement uses image matching algorithm to identify the same feature in two or more overlapped images. Tie points is responsible for orientating, tying all images in the strips and arranging all images similar as in the flight mission. Ground control points need to be established to register the images at the local coordinate system. In this study, ground control points were established using real time kinematic global positioning system (RTK-GPS). After both orientations were completed, then bundle adjustment operation can be performed. The accuracy of aerial triangulation was assessed. The photogrammetric results were discussed in the result and analysis section.

3 Result and Analysis

UAV captured image from a certain altitude and all acquired images were stored in general bands red, blue, green (RGB). All acquired images were processed using photogrammetric software to obtain the final photogrammetric results. Two primary results were produced; namely; digital elevation model (DEM) and digital orthophoto. DEM is generated after aerial triangulation process based on tie points measurement and ground control points. Digital orthophoto is generated from an individual orthoimages for each model in the photogrammetric block. The operation that combines all individual orthoimages is known as mosaic operation. Figure 3 shows the example of digital orthophoto and DEM that was produced from UAV images after photogrammetric processing.

Digital orthophoto shows the image of the study area from Nadir view and it is free from any distortions. In this study, the quality of the digital orthophoto from UAV images is similar to the other aerial platform such as manned aircraft. Figure 3 shows the digital orthophoto with ground resolution 8 cm which is very high resolution images. DEM provides the additional data on elevation value at the study area. Elevation value can be used to generate three dimensional models at the study area. 17 checkpoints have been established randomly at the study area in order to assess the quality of digital orthophoto and DEM from UAV images. All checkpoints were established using total station which is known as accurate survey technique. The results from total station survey were used as benchmark for this study. There were about 80 traverse stations that were established to complete observation at the study area. As mentioned in the introduction, three types of RTK GPS techniques have been conducted to fulfill the objective of this study. For the first technique, RTK GPS received adjustment from Iskandar Network for real

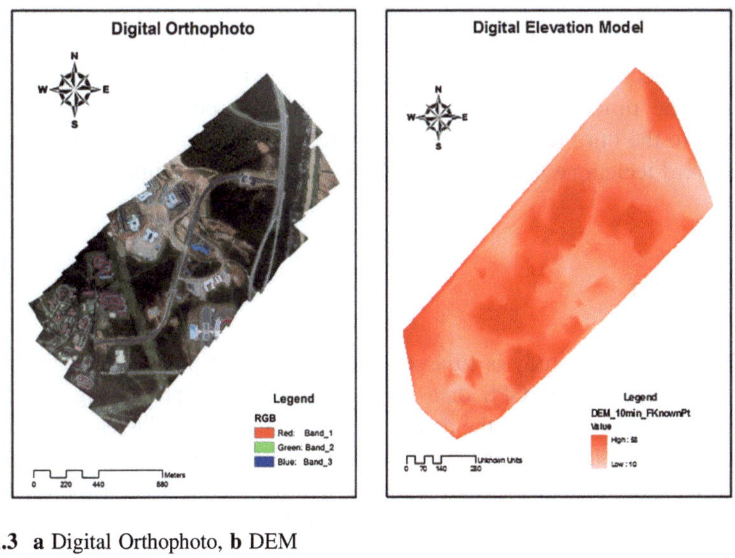

Fig. 1.3 **a** Digital Orthophoto, **b** DEM

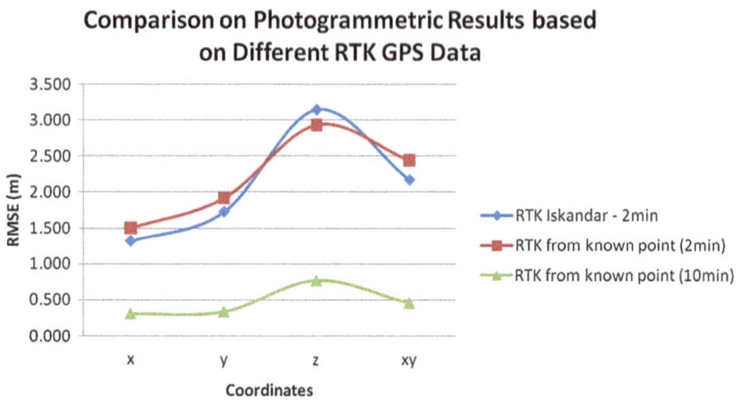

Fig. 1.4 Comparison on photogrammetric results based on different RTK GPS data

time adjustment during the control point observation. For the second technique, RTK GPS received adjustment from master station which was set up at the know GPS point and each point were observed for 2 min. Finally, for the third technique, RTK GPS received adjustment from master station has same result with the second technique except the observation time is 10 min. Figure 4 shows the comparison on photogrammetric results based on different RTK GPS data.

Based on Fig. 4 above; it was found that the accuracy between the first technique and the second techniques are almost similar. Specifically, the different x coordinates for the first and second technique is about ±18 cm, the different y coordinates for the first and second technique is about ±19 cm and the different z coordinates for the first and second technique is about ±22 cm. The range error for x, y and z coordinates are ±1.300 to ±3.1 m. However, the third technique gives different results from the first and second technique. All errors of the third technique were recorded below 1 m. Residual mean square error (RMSE) for x coordinates record ±31 cm, y coordinates record ±34 cm and z coordinates record ±77 cm. The difference between the first and the second technique is huge compared to the third technique. Based on the RMSE graph, it can be concluded that the third technique is the best RTK GPS data for UAV image processing.

4 Discussion

The objective of this study is to assess the accuracy of photogrammetric results based on different RTK GPS data during UAV images processing. Frequency distribution is able to pattern the data by grouping them into classes. In this study, frequency distribution is used to determine the pattern of residual error for each RTK GPS technique. Figure 5 shows the histogram of frequency distribution for the first, second and third RTK GPS technique based on the photogrammetric results.

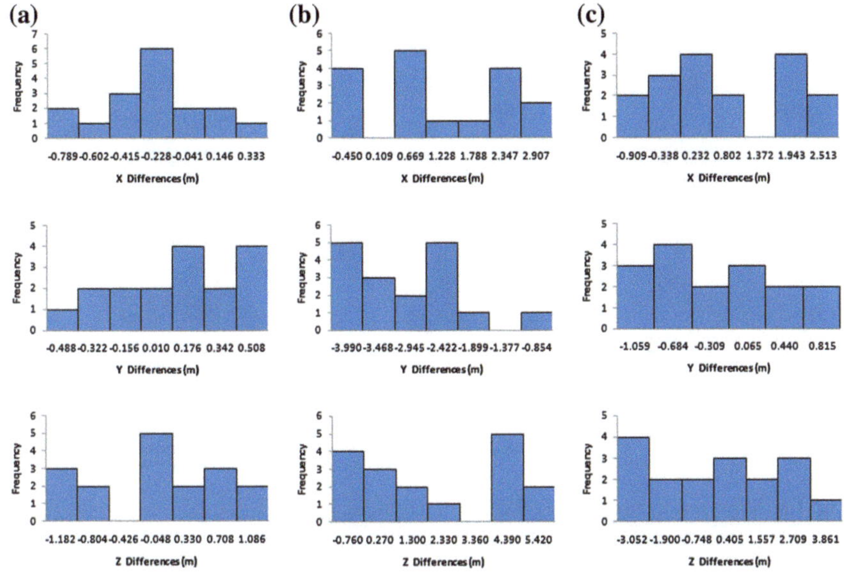

Fig. 1.5 Histogram, **a** RTK GPS from known point (10 min), **b** RTK GPS from known point (2 min), **c** RTK GPS Iskandar network (2 min)

17 checkpoints were established evenly at the study area in sequence to determine the precision of photogrammetric results based on different RTK GPS techniques. The frequency histograms were classified into seven classes which were equally divided based on residual error. Class relative frequency for RTK GPS from known point with ten minutes observation are 0.12, 0.06, 0.18, 0.35, 0.12, 0.12, 0.06 for easting coordinates, 0.06, 0.12, 0.12, 0.12, 0.24, 0.12, 0.24 for northing coordinates and 0.18, 0.12, 0, 0.29, 0.12, 0.18, 0.12 for height coordinates. Class relative frequency for RTK GPS from known point with two minutes observation are 0.24, 0, 0.29, 0.06, 0.06, 0.24, 0.12 for easting coordinate, 0.29, 0.18, 0.12, 0.29, 0.06, 0, 0.06 for northing coordinate and 0.24, 0.18, 0.12, 0.06, 0, 0.29, 0,12 for height coordinate. Class relative frequency for RTK GPS Iskandar Network are 0.12, 0.18, 0.24, 0.12, 0, 0.24, 0.12 for easting coordinate, 0.18, 0.24, 0.12, 0.18, 0.12, 0.12, 0.06 for northing coordinate and 0.24, 0.12, 0.12, 0.18, 0.12, 0.18, 0.06 for height coordinate. Easting coordinate for RTK GPS from known point with 10 min observation is more precise than RTK GPS from known point with 2 min observation and RTK GPS Iskandar Network. It is because easting coordinates of these technique shows bimodal histogram. Northing coordinate for Fig. 5a is skewed to the right while northing coordinates for Fig. 5b, c are skewed to the left. Height coordinates for all techniques portray bimodal histogram. Based on these histograms, it was found that RTK GPS from known point with 10 min observation gave the precise ground control points. It is because this technique received the real time correction data from the master station and the adjustment was calculated for 10 min for each point positioning. The other RTK GPS techniques described bimodal

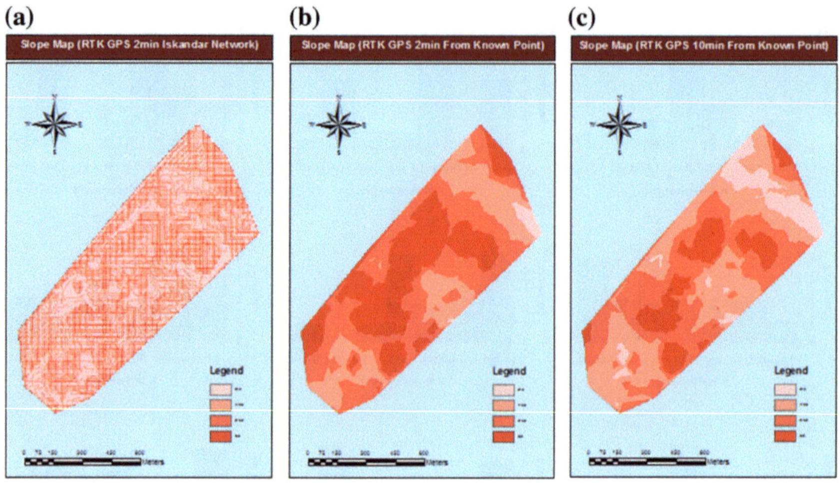

Fig. 1.6 Slope results for different RTK GPS techniques

histogram due to the inconsistent adjustment received during observation. In this study, we have generated three slope maps based on the different RTK GPS image processing (Fig. 6). Based on these results, it was found that, different RTK GPS data during photogrammetric image processing gives different attribute on slope database. Graphically, all RTK GPS techniques illustrate different slope attributes.

Figure 6 shows the slope map which can be viewed graphically by user for interpretation. The attribute for all slope maps were compared for numerical study. Figure 7 describes the pattern of slope attribute at the same area using different

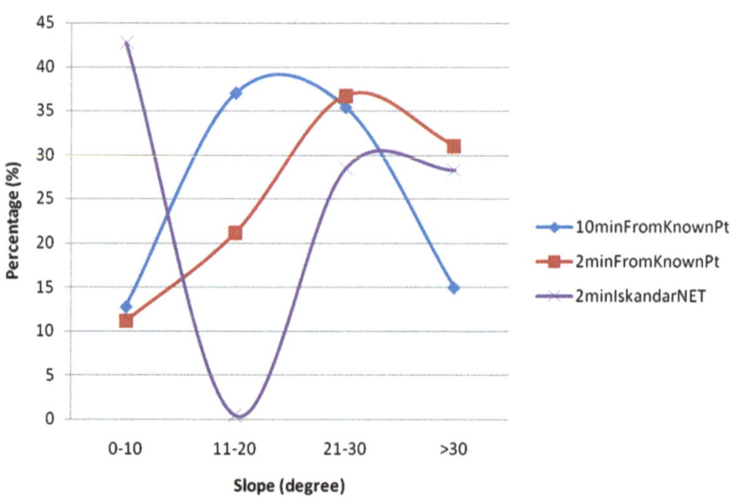

Fig. 1.7 Slope classification for different RTK GPS techniques

RTK GPS data during photogrammetric image processing. In numerical analysis, each results from different RTK GPS technique illustrates different attribute especially for slope class between 11 and 20° The percentage for slope class 0–10°is not significant between RTK GPS 10 and 2 min observation from known point but it is quite different for RTK GPS Iskandar Network.

5 Conclusion

This chapter discusses the photogrammetric results based on different RTK GPS data during photogrammetric image processing. There were three techniques being applied, namely, RTK GPS from known point with 10 min observation, RTK GPS from known point with 2 min observation and RTK GPS from Iskandar Network with 2 min observation. In conclusion, it was found that, RTK GPS from known point with 10 min observation gave the best result in term of accuracy and precision. It is because this technique received correction data from master station during observation with 10 min observation for one point positioning. UAV is one of the efficient equipment to obtain three dimensional models of the area of interest especially at the slope area. In the future, the comparison between engineering surveying using total station and UAV will be explored to determine the cost, accuracy and time taken for one mapping project.

Acknowledgments Faculty of Architecture, Planning and Surveying Universiti Teknologi MARA (UiTM) and Faculty of Geoinformation & Real Estate, Universiti Teknologi Malaysia (UTM) are greatly acknowledged. The authors would also like to thank the Sustainability Research Alliance, Universiti Teknologi Malaysia for providing the fund to enable this study to be carried out.

References

Abellán A, Calvet J, Vilaplana J, Blanchard J (2010) Detection and spatial prediction of rockfalls by means of terrestrial laser scanner monitoring. Geomorphology 119(3–4):162–171

Al-Bayari O, Saleh B (2007) Geodetic monitoring of a landslide using conventional surveys and GPS. Surv Rev 39(305):352–356

Barber D, Mills J, Bryan P (2001) Laser scanning and photogrammetry: 21st century metrology. Paper presented at the 18th International Symposium CIPA, Potsdam, Germany

Cardenal J, Delgado J, Mata E, González-Díez A, Remondo J, de Terán J, Francés E, Salas L, Bonachea J, Olague I, Felicisimo A, Chung Ch J, Fabbri A, Soares A (2006) The use of digital photogrammetry techniques in landslide instability. In: Cruz G, Sanso, (eds) Geodetic deformation monitoring: from geophysical to geodetic roles, IAG Springer Series, Berlin, pp 259–264

Gili J, Corominas J, Rius J (2000) Using global positioning system techniques in landslide monitoring. Sci Direct Eng Geo 55:167–192

Hai YC, Yong CC, Yang QC (2010) Autopilots for small unmanned aerial vehicles: a survey. Int J Con Autom Sys 8(1):36–44

Janssen V, Grinter T, Roberts C (2011) Can RTK GPS be used to improve cadastral infrastructure. Eng J 5(1):43–54

Johan K, Christophe D, Pascal A, Pierre P, Marion J, Eric V (2010) Application of a terrestrial laser scanner (TLS) to the study of the Séchilienne Landslide (Isère, France). Rem Sens 2:2785–2802

Lei Y, Zhiyang G, Yini D (eds) (2009) A UAV remote sensing system: design and tests. In: Li D, Shan J, Gong J (eds) Geospatial technology for Earth observation data, Springer, New York

Li J, Ouyang Z, Wei X, Zhou H (2010) Application of reflectorless total station measure technique in landslide deformation monitoring. The Chinese Journal of Geological Hazard and Control, 1, 028.

Montané JM, Torres R (2006) Accuracy assessment of Lidar Saltmarsh topographic data using RTK GPS. Photogramm Eng Rem Sens 72(8):961–967

Moss J (2000) Using the global positioning system to monitor dynamic ground deformation networks on potentially active landslides. Int J Appl Earth Obs Geo 2(1):24–32

Pesci A, Baldi P, Bedin A, Casula G, Cenni N, Fabris M, Loddo F, Mora P, Bacchetti M (2005) Digital elevation models for landslide evolution monitoring: application on two areas located in the Reno river valley (Italy). Ann Geophys 47:1339–1353

Pocsai A, Zámolyi A, Székely B, Molnár G, Roncat A, Drexel P (2011) Change detection on the Doren landslide, using geological field measurements and laser-scanned data (Vorarlberg, Austria) Geophys Res Abstr 13: EGU2011-938

Squarzoni C, Delacourt C, Allemand P (2005) Differential single-frequency GPS monitoring of the La Valette landslide (French Alps). Eng Geo 79:215–229

Teza G, Galgaro A, Zaltron N, Genevois R (2007) Terrestrial laser scanner to detect landslide displacement fields: a new approach. Int J Rem Sens 28(16):3425–3446

Teza G, Pesci A, Genevois R, Galgaro A (2008) Characterization of landslide ground surface kinematics from terrestrial laser scanning and strain field computation. Geomorphology 97(3–4):424–437

Tahar KN, Ahmad A, Wan Mohd Akib WAA (2011) Unmanned aerial vehicle technology for low cost landslide mapping. Paper presented at the 11th South East Asian Survey Congress and 13th International Surveyors' Congress, PWTC, Kuala Lumpur, pp 22–24

Tahar KN, Ahmad A (2012) A simulation study on the capabilities of rotor wing unmanned aerial vehicle in aerial terrain mapping. Int J Phy Sci 7(8):1300–1306. doi:10.5897/IJPS11.969

Tarchi D, Casagli N, Fanti R, Leva D, Luzi G, Pasuto A, Pieraccini M, Silvano S (2003) Landslide monitoring by using ground-based SAR interferometry: an example of application to the Tessina landslide in Italy. Eng Geo 68:15–30

Yue D, Wang J, Zhou J, Chen X, Ren H (2010) Monitoring slope deformation using a 3-D laser image scanning system: a case study. Mining Sci Tech (China) 20(6):898–903

Towards Enhancing Geometry Textures of 3D City Elements

Behnam Alizadehashrafi and Alias Abdul Rahman

Abstract In this research, different texture enhancement methods were investigated and proposed. Rectified perpendicular symmetric photos of 3D objects such as buildings and trees were used to generate the texture for the 3D models. Texture enhancement methods such as higher radiometric adjustment for raised surfaces and Dynamic Pulse Function (DPF) were utilized based on the form of the façade and operators' observation along with tessellation method. The result of DPF compared with the photorealistic texturing in aspects, quality and data size. The problem of missing faces in a polyhedron and missing object within a CityGML viewer has been solved and described in this chapter. The chapter also describes the development of External code-Lists for objects of the study area which is a starting point for Malaysian Spatial Data Infrastructure (MySDI).

Keywords Dynamic pulse function · Texture enhancement · Symmetric transformation · Semantic modeling · CityGML · CityServer3D

1 Introduction

Unlike computer graphics, visualization is not the only target of GIS. Semantic information and database are vital in GIS for analyzing the data and 3D object characteristics within a virtual urban area. Web-based applications such as Google Earth and Microsoft Virtual Earth are suitable for visualization purpose but not for

B. Alizadehashrafi (✉) · A. Abdul Rahman
3D GIS Research Lab, Faculty of Geoinformation and Real Estate,
Universiti Teknologi Malaysia, Johor Bahru, Malaysia
e-mail: behnam@utm.my

A. Abdul Rahman
e-mail: alias@utm.my

A. Abdul Rahman et al. (eds.), *Developments in Multidimensional Spatial Data Models*, Lecture Notes in Geoinformation and Cartography,
DOI: 10.1007/978-3-642-36379-5_9, © Springer-Verlag Berlin Heidelberg 2013

Spatial Data Infrastructure (SDI) and database. The concept is to address 3D modeling of any objects within the city along with semantic database such as terrain, road network, building wireframes or solid models and street furniture. Pictometry is a new highly automated 3D modeling technique and can be used for lots of different applications but there are some serious texturing problems in this technique such as shadow, reflection, disturbing objects, fog, pollution and distortions.

2 Dynamic Pulse Function

Coors and Zipf (2007), defined an algorithm for creating the texture for façade on the mobile devices based on limited number of layers using pulse function (PF) on J2ME platform. The priority of the door layer is higher than window layer and it can be placed on left, middle or right side of the façade and the window cannot be placed on the door. Alizadehashrafi et al. (2009) implemented the same concept of aforementioned PF method to generate the façade image based on unlimited number of layers but on a fixed output frame which was not proportional to real geometry. This method caused deformation and distortion while stretching or shrinking the generated façade image on the geometry. This system could create the output image file in a very high quality and small data size along with XML (eXtensible Markup Language) schema file for different layers such as walls, doors and windows but the quality reduced while mapping on the geometry because of distortion. The output image file was even smaller than the textures which were used to create the image of the façade. 3D textured model in VRML was the final product in LOD2 (Level of Detail 2). Virtual Reality Modeling Language (VRML) does not support procedural texturing and Pulse Function (PF) so the final pre-processed procedural textures based on Dynamic Pulse Function (DPF) were mapped on the 3D buildings wireframe in VRML manually for Stuttgart downtown and employed by Bogdahn and Coors (2011).

Texture deformation while mapping the square output image on rectangle geometry, is the main problem of PF method along with meaningless database of parameters within the XML schema file which are not proportional to reality. These leads to lack of quality and texture deformation in the 3D models along with XML database such as size and position of windows and doors which may not be useful for query in the near future. To deal with this problem, DPF comes in handy and can be used for semantic and geometric database such as name of layer, its position, size and cluster and so on. As the output frame is proportional to real geometry in DPF, based on height and width of the façade it is possible to query the XML database and convert them to units such as meter automatically. In this technique, the perpendicular terrestrial photo from the façade is rectified via projective transformation based on the frame in constrain proportion to real geometry. The rectified photos of the whole façade which is not suitable for texturing but adequate for measuring parameters, must be down-sampled in constrain proportion to real geometry before measuring process. For instance height and width of the resized and rectified photo which is proportional to real geometry can be employed for generating the output frame for procedural texturing based on real textures of windows, doors and walls. The height and width of windows, doors, horizontal and vertical distance between them in image coordinate system can be measured via number of pixels. The system can use these parameters and texture file names and file paths to create the façade semi-automatically. The dimension of output image file is proportional to the real geometry to prevent distortion while projection. To avoid leaning geometry the textures of windows, doors and etc., should be cropped and rectified from perpendicular photos and enhanced, so that they can be used in the program to create the whole façade. Texture enhancement should be done in advance such as removing disturbing objects, exposure settings, left-right up-down transformation, and so on. Figure 1 illustrates the process of texture generation based on DPF. Figure 2a, b are original side view photo and rectified façade respectively and Fig. 2c represents the projection of the image on the virtual 3D geometry and finally Fig. 2d shows the output procedural façade image based on DPF nearly 17 KB in a very high quality including semantic database on XML schema file. Figure 3 illustrates the enhanced textures which are employed for DPF method. In fact, quality, small data size and scale along with semantic database for each façade are the advantages of DPF method for texturing 3D models and querying the parameters and positions of the object on the façade.

Parameters of windows and doors with respect to upper left corner of the rectified façade or image coordinate system can be measured via programming in Matlab after rectification and projective transformation (see Fig. 4).

These parameters can be used for generating XML schema file based on DPF (see Fig. 5).

Due to lots of raised or deep geometries in some buildings, it is not possible to assume the whole façade of the building as a unique façade. To deal with this problem, the operator should divide the façade heuristically and generate the

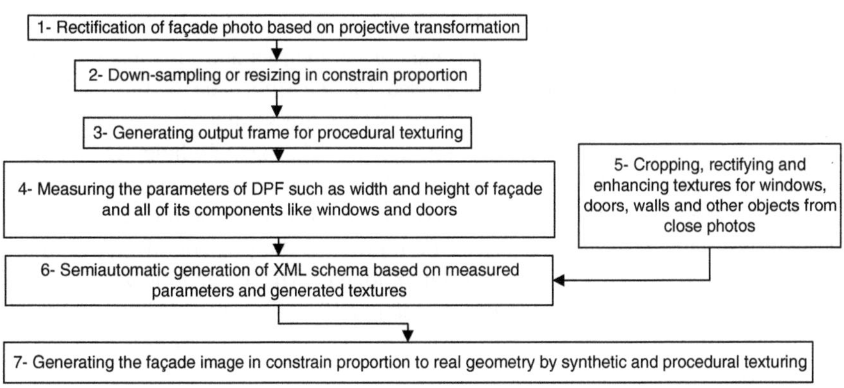

Fig. 1 The flowchart for dynamic pulse function

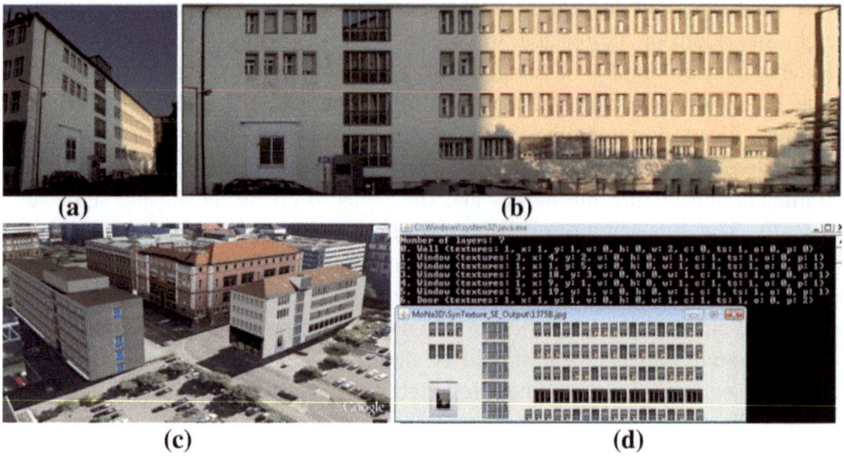

Fig. 2 **a** Original side view photo (4 MB). **b** Rectified photo (7 MB). **c** 3D model on google earth. **d** Generated photo (17 KB in a very high quality)

Fig. 3 One pixel is used for the wall texture and three random textures are used for three window layers to increase level of realism (LoR)

texture for each part separately. DPF can be used for each subsection to tackle problems such as lack of quality and leaning geometries and also using efficient storage. The generated texture can be used for the right side of the façade via flipping the texture in left–right direction, as they are symmetrical textures. This can reduce the data size of 3D model significantly. The size of whole 3D model for each building in Kolej Perdana is around 133 KB. Figure 6a, b are java code for creating the classes for DPF and JavaScript code for generating the XML schema file based on DPF respectively. Figure 6c represents the procedural façade image (15 KB) based on DPF by running aforementioned XML schema file via generated java classes. Figure 6d illustrates the cropped and rectified photo of the façade for measuring purpose and finally Fig. 6e is the side view photo from whole façade before cropping and dividing.

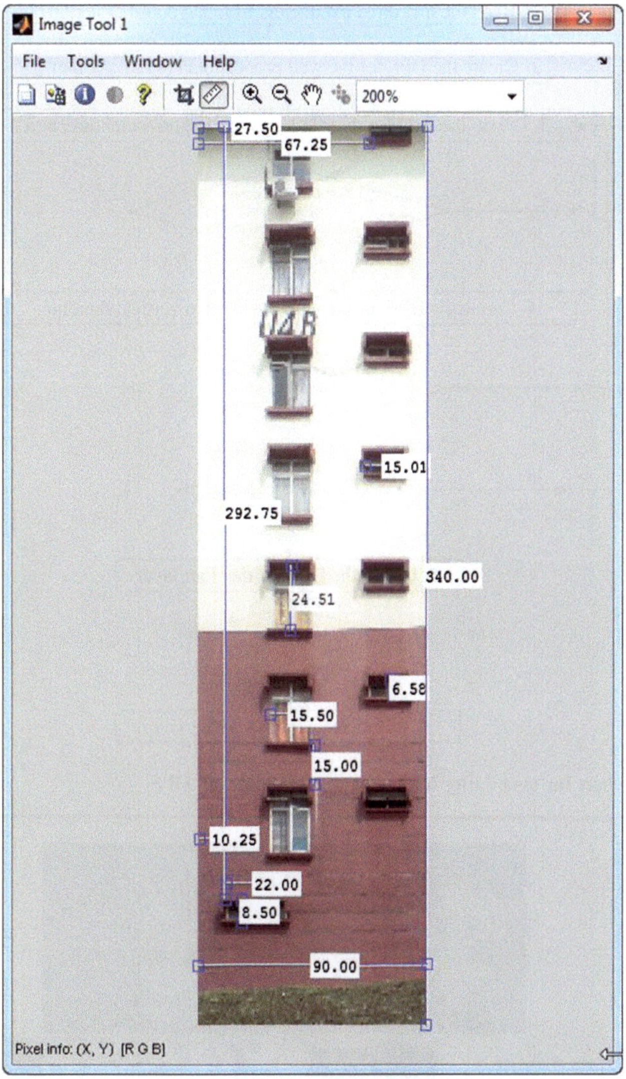

Fig. 4 Measuring all the required parameters for DPF via programming in Matlab

3 Increasing the Quality of the Pictometry 3D Models by DPF

Pictometry oblique photos with roughly 40° angle are not good enough for texturing 3D objects. In case of having database of textures for windows, doors and walls from a newly constructed urban area, these textures can be used based on DPF in order to regenerate the textures. The pictometry applications can be

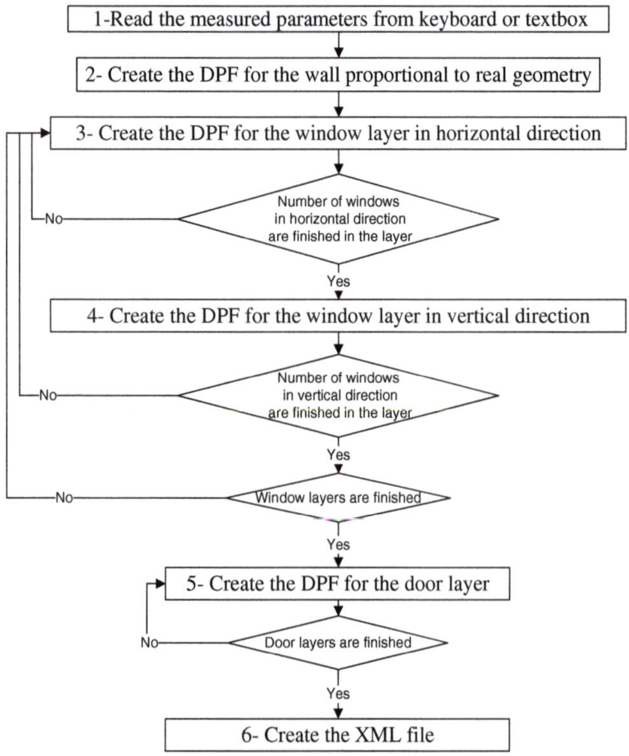

Fig. 5 Flowchart for generating XML schema file based on DPF

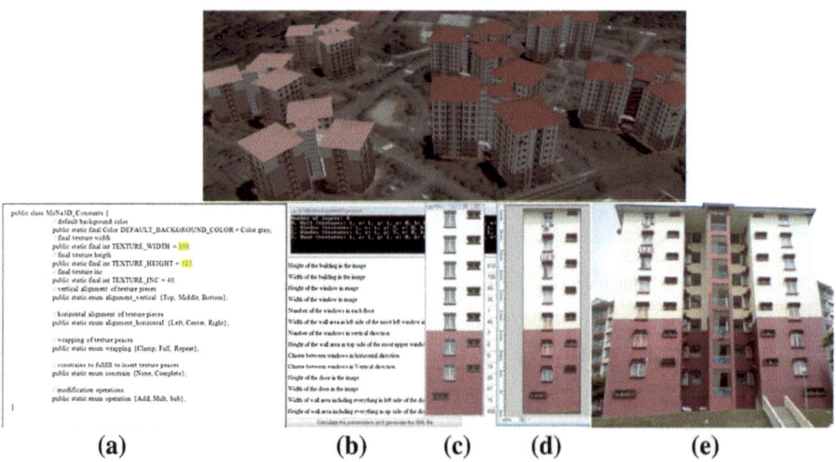

Fig. 6 The process of generating the façade via DPF and visualization in Google Earth. (*Upper image* is Kolej Perdana, UTM Campus, Skudai, Malaysia in Google Earth)

employed for measuring width and height of the façade, windows, doors and other necessary parameters. These parameters along with existing database of textures can be used in DPF in order to regenerate the textures for the pictometry 3D models in a very high quality and small data size. Figure 7a, b are bird eye view photo and rectified façade respectively and Fig. 7c represents the projections of the procedural photo on the 3D geometry via visualization in Google Earth and finally Fig. 7d illustrates the process of generating the texture from oblique pictometry photo by employing DPF and existing database of window and wall textures.

4 Features of JavaScript Program for Generating XML File

In the JavaScript program, importing and selecting window and door layers are not compulsory but wall texture. If there is nothing in the textbox for the window and door layers, the program can recognize automatically and no layer will be created in the XML file for door and window. In addition, unlimited number of random textures can be used for each window layer to increase Level of Realism (LoR)

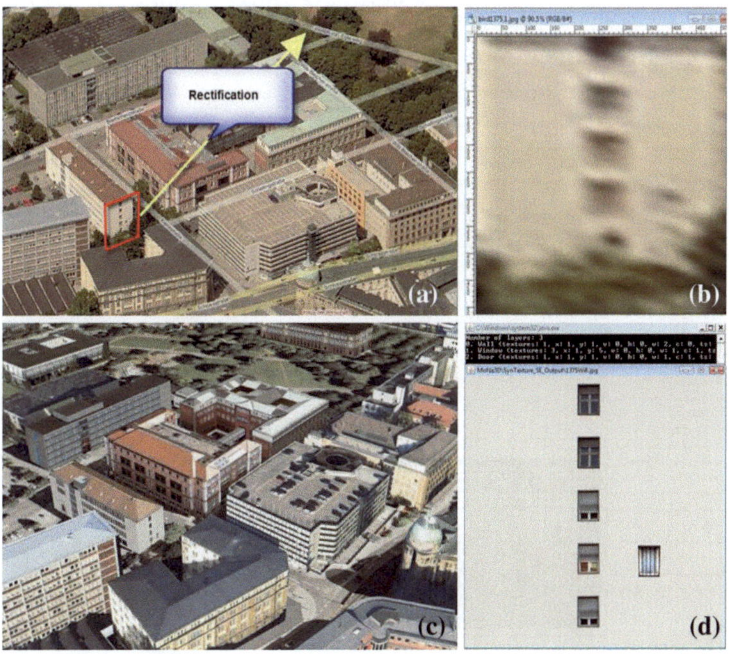

Fig. 7 **a** Original oblique pictometry photo from microsoft virtual earth. **b** Cropped and rectified façade photo from original photo. **c** Visualization of the 3D models based on DPF in Google Earth. **d** Procedural photo generated based on existing database and DPF

(see Fig. 8). The background layer is using and tessellating the texture based on dimension of the texture and DPF frame size. If the height of the background texture and frame are equal, the texture tessellates horizontally to generate the background (see Fig. 9). In case of equal width of background texture and frame, it repeats vertically to cover the background. For instance, by tiling one pixel in both, horizontal and vertical direction the background wall can be created.

5 Problems and Limitations for Dynamic Pulse Function

Some of the building façade cannot be generated by DPF because of strange and wired form of background or shape and geometry. The concept behind DPF is to repeat one or more small sample of texture in a very high quality to create whole façade along with geometric information. This can be employed for uniform facades or symmetrically uniform facades. For instance, it is very difficult to use

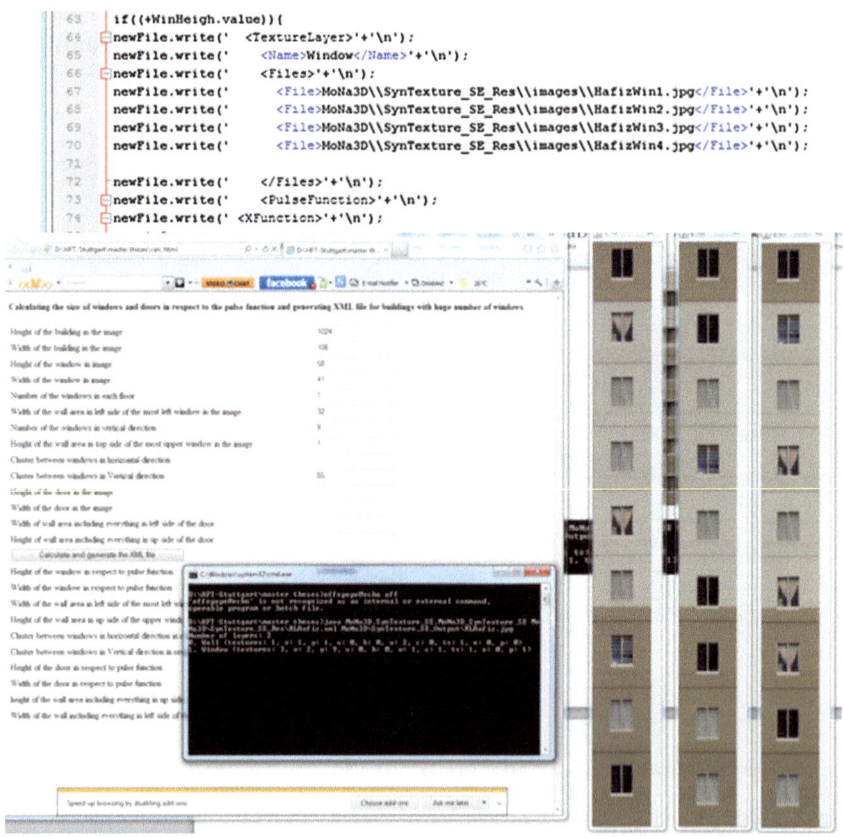

Fig. 8 JavaScript program uses random textures for window layer to increase level of realism

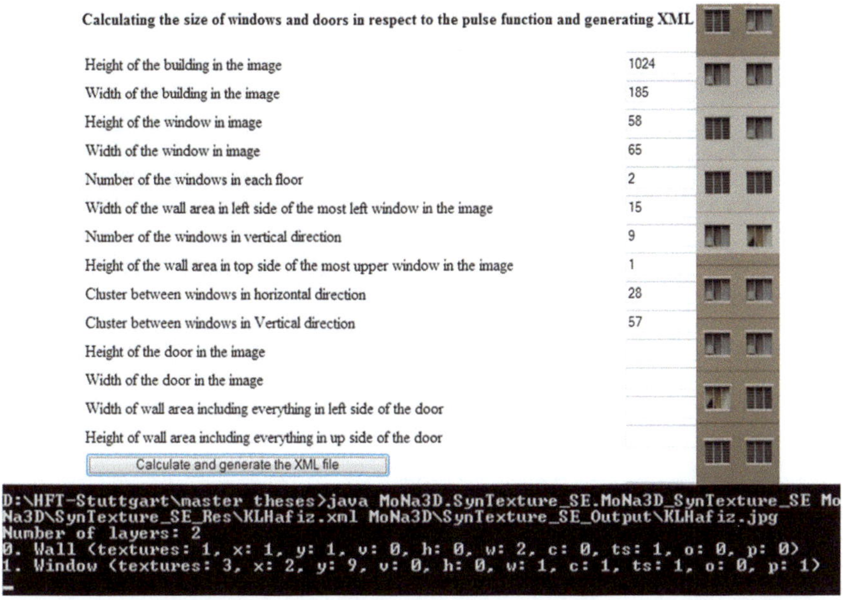

Fig. 9 JavaScript program creates the XML file for the façade and Java graphics generates the façade using two layers for wall and window

Fig. 10 Samples that cannot be textured by dynamic pulse function

this technique for the left middle constructions in Fig. 10 because of abnormal geometries and right one because of weird background texture.

In addition to aforementioned façade textures, DPF cannot be used for some complex geometry such as heritage, historical and religious landmarks and their roof shape which is mostly in domical form. The shape of the window is not an important issue in this method as PNG file can be employed as a window texture that supports transparency and it is easier to edit than GIF and also smaller in terms of data size. In fact, the shape of the image is in rectangle form in computer graphics. The picture elements are usually indexed from upper left corner to lower right corner, but it is possible to make some parts of the image transparent as needed to create the texture for the window (see Fig. 11). The final textured 3D model is illustrated in Fig. 12 in a very high quality and without disturbing object

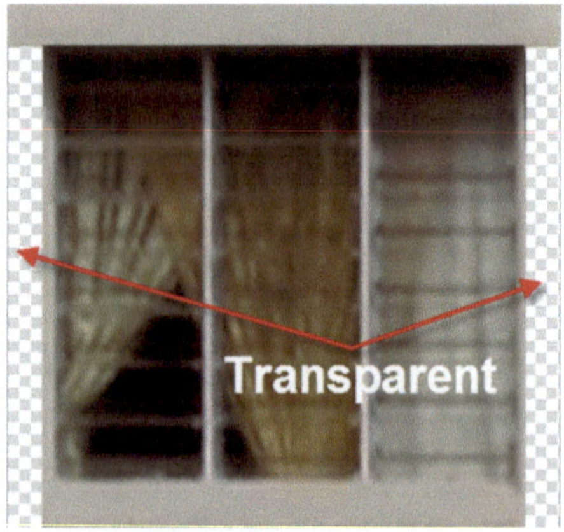

Fig. 11 The image is rectangle but the window texture is not rectangle and has transparent parts

Fig. 12 Textured via DPF

and leaning geometries along with semantic definitions for window layer and wall layer in XML file.

6 Comparison Between Photorealistic Texturing and DPF

The photo of the façade, taken by wide angle lens, should be rectified based on projective transformation or texture generation process and mapped on the 3D wireframe. It is better not to reduce the dimension of the rectified photo as it can cause deformation and lack of quality. Disturbing objects such as air-conditioners, hanging clothes, pedestrians, parked cars and so on, can be edited and removed manually by photo editing software such as Photoshop but it is time consuming. In this method, it is not possible to solve the problem of leaning windows especially for upper windows on the rectified photo taken from ground level (See Fig. 13).

Fig. 13 Textured via photorealistic method

The result shows that the final façade image based on DPF has smaller data size not only than photorealistic image but also than its own windows textures. Even at the same dimension, the data size of final 3D model based on DPF is almost thirty-five percent less than data size of photorealistic 3D model. It shows that the data size of the generated texture is one fifth of the rectified photo. In addition semantic information of components of the façade such as window layer can be queried from the database in XML schema.

7 Tiling and Tessellating

In Putrajaya area, data collection has been done by 10.2 mega pixel Nikon D60 camera. The image files are prohibitively large. Although it is possible to overcome this problem by reducing the resolution of the image but such down sampling can cause lack of quality and loss of information. The textures which are tessellating vertically should be transformed symmetrically in up-down direction and those which are tessellating horizontally should be symmetrically transformed in left–right direction. This can create a horizontal or vertical symmetric texture. In case of tiling the texture both vertically and horizontally, the combination of aforementioned methods can be used. The algorithm and concept is very simple and can be coded in any programming language or even photo editing applications to produce very high quality textures. This technique applied to the floor of Putra Mosque (see Figs. 14, 15).

8 Texturing Sophisticated Domical Geometries

In some of the complex geometries such as the domical geometries, perpendicular photos were taken from the dome and after rectifying symmetrically; the image is imported to the SketchUp as an image or standalone object vertically. Half of the

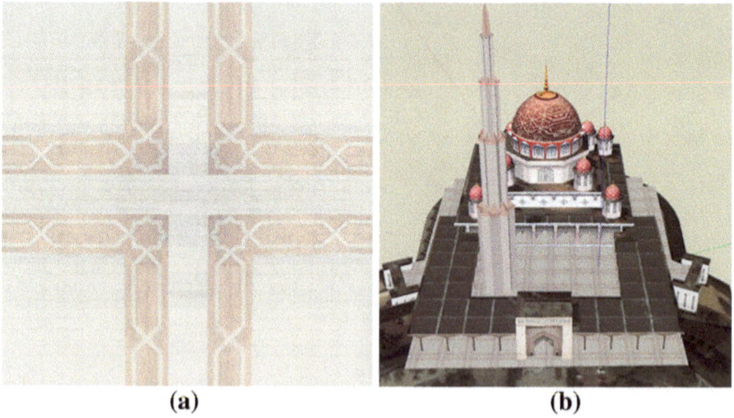

Fig. 14 a The created texture for the floor based on rectification and symmetric transformation. b Better visualization and higher LoR can be seen in the model after tessellating

Fig. 15 Illustration of the textures which are created based on symmetrical left-right transformation to increase LoR. a Putrajaya holdling department. b The court of justice. c The Ministry of Finance

dome geometry is sketched in a planer surface of the image. A void or hollow circle is sketched as a footprint of the dome based on its diameter. Available tools are used to generate the geometry of the dome. Finally the texture projected from the image on the dome geometry and the dome defined as a component in order to be moved to its proper position (See Figs. 16, 17).

Fig. 16 Modeling dome of the Putra and Sultan Mizan mosques based on rectified and symmetric photo taken from the same level in Putrajaya

Fig. 17 Using higher exposure settings to represent raised geometry for Putrajaya Prime Minister Department

9 Texturing Based on Different Exposure Setting and Alpha Channel

From the architectural point of view, the geometries which are closer to the camera should be brighter and those which are farther should be darker. It is possible to create a very light-weighted geometry in LOD 2 for web-based applications such as Google Earth, Microsoft Virtual Earth and mobile navigators are based on this concept to reduce the geometries and textures accordingly. To increase LoR, the geometry with lots of raised columns represented as flat rectangle geometry and the brightness of the columns are increased or for the deeper parts, the brightness decreased. This is a trick, which can be used for reducing the geometry and number of texture files and increase LoR as well. Figure 17 shows the concept clearly. In the texture of the façade, the RGB pixel values are increased for the column in left side of the texture to represent raised geometry and then the texture left–right transformed symmetrically and finally mapped on the rectangle geometry.

Some statues in Stuttgart in front of New Castle, in front of the church, within the garden in front of Stuttgart University of Applied Science, annexes between the faculties along with cement fences in University Technology Malaysia (UTM) and

Fig. 18 3D models of statues, annexes between the departments and fences based on Alpha Channel

trees based on the concept of alpha channel were modeled (see Fig. 18). CityServer3D for semantic database were used for building information in Putrajaya. 3D models of statues, annexes between the departments and fences were created based on Alpha Channel and projected on the rectangular polyhedron geometry. Some of lamp posts and trees were created on two perpendicular rectangular geometries with the thickness of 1 mm, intersecting each other from the middle at 90° angle or three of them intersecting at 60° angle. The main reason for using rectangular polyhedron with 1 mm thickness in SketchUp before converting to CityGML, is the problem of invisible standalone graphs without geometry in CityGML. For the purpose of 3D model in Google Earth or SketchUp, just the standalone PNG graph file with its transparent part instead of rectangular polyhedron with 1 mm thickness geometry can be used without any invisibility problem.

Palm or coconut trees with the uniform leaves were designed based on repeating one 3D leaf model on the trunk in different scales, positions and rotations (based on 7 parameters, 3 for position, 3 for rotation and 1 for scale for each leaf on the trunk) to increase LoR (See Fig. 19).

10 Healing of Invisible Faces in CityGML

Mostly the models which are designed in SketchUp without any texture problems may have missing faces or solid non-textured faces in CityGML after conversion in any CityGML viewer such as KMZ viewer, LandXplorer and CityServer3D because of non-oriented faces. An automated 3D healing model has been proposed and started by, in Stuttgart University of Applied Science in order to solve this problem. The system receives original model and based on a validation and healing plan it will generate correct model automatically or create correction recommendation for the operator to heal the 3D model based on operator's supervision. Finally the system creates the quality and healing report. The complexity of this algorithm is O (n!) and can only be computed for a small number of points in a 3D model.

Fig. 19 Illustration of designing a palm tree via transparency and converting it from SKP to VRML (BitManagment) and CityGML (LandXplorer) including 892 faces, 653 edges and 2 texture files

The list of nodes for each polygon, facing the camera position in CityGML, must be counter clockwise so that it can be visible in any CityGML visualizer. The solution for this problem proposed by the author to make all the faces counter clockwise before converting to CityGML in SketcuUp via Orient or Inverse options. By checking the 3D model in Monochrome mode, the operator can find those faces which are in grey colour (not oriented) rather than white colour (oriented). These faces can be reversed one by one or whole model can be healed via Orient option before converting to CityGML without using the algorithm which is provided by Bogdahn and Coors (2010). Two faces on the roof of C26 in UTM are in grey color in monochrome mode in Fig. 20a (not oriented) and in normal mode they are visible in SketchUp environment. After converting to CityGML these faces are invisible via any visualizer like LandXplorer, from outside as they are not oriented (see in Fig. 20b). This problem was solved in Fig. 20c after reversing and retexturing in SketchUp before converting to CityGML, as it can be seen in LandXplorer (see Fig. 20).

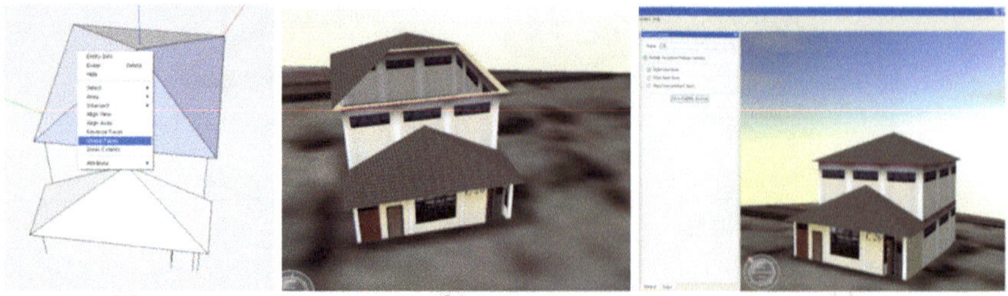

(a) (b) (c)

Fig. 20 Solution for invisible non oriented faces in CityGML

11 External Code Lists and Semantic Modeling in CityServer3D

In order to represent the object attributes of the city, having an enumerative range of values is necessary and the concept of dictionaries as provided by GML should be used. The values are defined in a file called CityGML_ExternalCodeLists.xml, which comes with the CityGML schema document, but is not a normative part of this schema, since it may be modified, augmented, or replaced, by users or other communities (Groger et al 2009). The actual values in the file CityGML_ExternalCodeLists.xml are a suggestion of the SIG 3D (Special Interest Group 3D). In the process of designing semantic and thematic information for 3D city objects, these predefined attributes are necessary to assign to these objects which can be standardized for each area and country such as Malaysia. Attributes are employed to classify the objects and recognize them via queries or clicking and selecting the objects within 3D virtual environment of CityServer3D. It is similar to indexing system and a unique value relates to each attribute within a single file. These values are defined according to the attributes' name. The files are placed in a directory, which is called code lists. The code lists for the predefined objects in Germany can be extracted from the CityGML schema and the data can be changed, modified, enlarged and replaced according to requirement of the data and user's needs. These code lists are initially equipped by SIG 3D and defined by Open Geospatial Consortium (OGC) in CityGML (Gröger et al 2008).

The external code list file defines attribute values and assigns a unique identifier to each value. In a CityGML instance document, an attribute value is denoted by an identifier of a value, not by the value itself. These identifiers are known terms for the operators and users. Thus, printing or typing errors are avoided and it is ensured that the same concept is denoted the same way, by the same identifier and not by two different terms with identical meaning. This is why the use of code lists facilitates semantic and syntactic interoperability, since they define common terms within an information community. Furthermore, the dictionary concept enables more than one term to be assigned to the same dictionary entry, thus the same concept may be explained in different languages. To differentiate between the languages, code spaces are used.

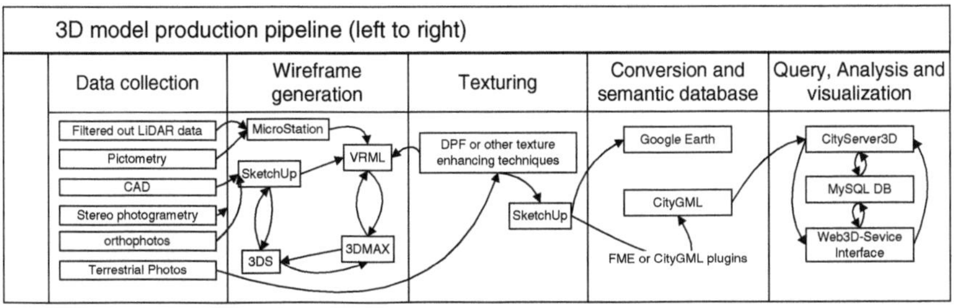

Fig. 21 3D model production pipeline for visualization, analysis and semantic modeling for CityServer3D

Fig. 22 DTM produced using Google Earth terrain with the grid size of 76 m

'Definition Collection' element is used in the dictionary concept for representing the values of an attribute, where each value is given by a Definition entry. In CityGML a definition entry is identified by the name element, which is qualified by the SIG 3D code space. The unqualified name element represents the value of the attribute and an optional description defines the value. CityGML does not use GML identifiers (gml:id) to link to attribute values, since IDs are restricted syntactically, and must be globally unique, which is not feasible for code lists.

All the required code lists are not available in CityGML external code lists though it covers most of the city objects. Some code lists are proposed according to the suitability for Malaysia especially in vegetation types such as species type because the vegetation type of each country is different according to climate conditions. In addition to the proposed code lists for the trees and plant species in Malaysia some code lists were proposed for roof types of the buildings.

Figure 21 illustrates the process of 3D model production pipeline from left to right. In this research only the CAD data and terrestrial photo where used to

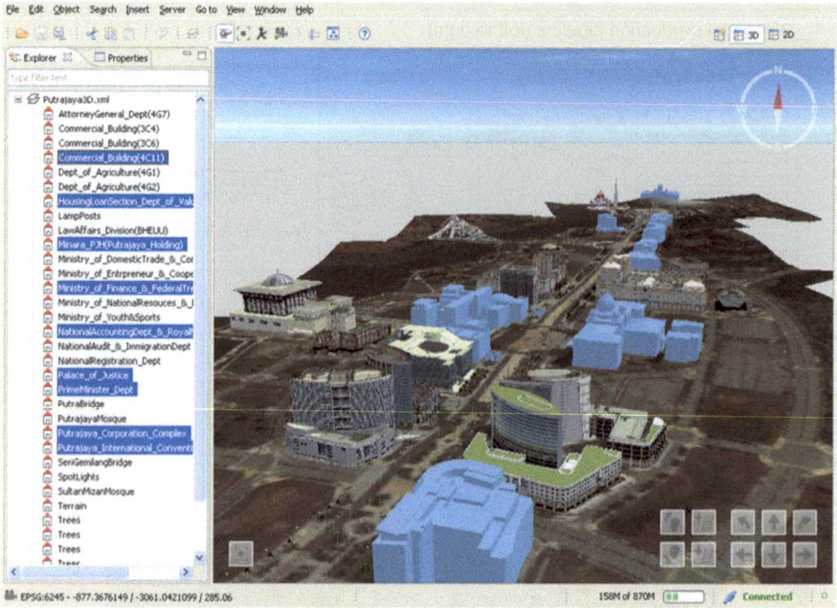

Fig. 23 Visualization of Putrajaya3D including semantic database via client system's Admin-Tool from CityServer3D

generate the 3D models along with aforementioned texture enhancement techniques. The 3D conversion process is shown in column 2, 3, 4 and 5. Semantic information and code-lists are added in column 5.

The terrain and Ortho-photo of an area from Putrajaya Prime Minister Department (PPMD) to Putrajaya International Conventional Centre (PICC) about 5 km by one kilometer created using Google terrain with the grid size of 76 m (see Fig. 22). 25 buildings, 4 bridges and all the street furniture such as spot lights, lamp posts, trees, and traffic lights were modeled and textured by aforementioned methods in this area along with semantic data in CityServer3D (see Fig. 23) Alizadehashrafi and Rahman (2011).

12 Summary

2D existing highly detailed CAD data with significant number of layers and rectified vertical symmetric photos of intricate constructions are used to generate the 3D models of Putrajaya in CityGML and CityServer3D. Up on request some of the 3D models were designed and textured in LOD2 and some in LOD3 and LOD4 with higher LoR. Texture enhancement methods such as texture left–right or up-down symmetrical transformation and higher radiometric adjustment for raised surfaces and DPF for procedural texturing targeting the high quality along with small data size were used and explained. These texturing methods can be done based on the situation of the façade and observation of the operators. The problem of missing faces of polyhedrons and missing objects have been solved and described based on inversing and orienting the faces or designing geometries before texture projection for trees and lamp posts in SketchUp before exporting to CityGML. Semantic information and external code lists were defined and added to the database on CityServer3D for Putrajaya area (Munankarmi 2011).

Acknowledgments We would like to convey our deepest acknowledgement firstly to Universiti Teknologi Malaysia (UTM) and our sincere appreciations to the Malaysian International Scholarship (MIS) for enabling us to carry out this research project.

References

Alizadehashrafi B, Rahman AA (2011) CAD-based 3D semantic modeling of Putrajaya. In: Proceedings of the Joint ISPRS Workshop on 3D city modelling & applications and the 6th 3D GeoInfo conference

Alizadehashrafi B, Rahman AA, Coors V, Schulz T (2009) 3D navigation systems based on synthetic texturing, vol 56. Wscg, Poster Proceedings, pp 21–26

Bogdahn J, Coors V (2010) Towards an automated healing of 3D urban models In: 5th international 3D GeoInfo conference, Nov 3–4, Berlin, Germany, XXXVIII-4/W15, pp 13–17

Bogdahn J, Coors V (2011) Mobile pedestrian navigation using 3D city models and procedural façade textures. 15

Coors V, Zipf A (2007) MONA 3D—Mobile navigation using 3D City models

Groger G, Kolbe TH, Czerwinski A (2009) Candidate openGIS cityGML implementation specification

Gröger G, Kolbe T, Czerwinski A, Nagel N (2008) OpenGIS® City Geography Markup Language (CityGML) Encoding standard. [Online]. [Accessed]

Munankarmi M (2011) 3D semantic city modelling of Putrajaya area, Malaysia Master of Science, Stuttgart University of Applied Sciences

Examination of Compatibility Between the Croatian Land Administration System and LADM

Nikola Vučić, Miodrag Roić and Zdravko Kapović

Abstract Croatian land administration system faces a great crossroad because it demands for a single maintenance solution of digital cadastral plans converted from an analogue to a digital form a few years back. Joint Information System of Land Books and Cadastre is currently under development. For the cadastral system transformation compliance with international standards is very important. Furthermore, introduction of the new geodetic datum and the new cartographic projections adopted in 2004 is still in its implementation phase. At the same time, Croatia is in the process of being admitted into full European Union membership (expected accession date is July 1, 2013). This chapter examines compatibility of the Croatian land administration system with the Land Administration Domain Model (LADM). Possibility of LADM adoption in the Republic of Croatia is evident from comparison of LADM literature and the current situation of the Croatian land administration system. This chapter proposes a national LADM profile for the Republic of Croatia. Interconnections between classes of the Croatian land administration system and linkage possibilities of these classes with LADM are presented through relevant examples.

N. Vučić (✉) · M. Roić · Z. Kapović
University of Zagreb, Faculty of Geodesy, Kačićeva 26, 10000 Zagreb, Croatia
e-mail: nvucic@geof.hr

M. Roić
e-mail: mroic@geof.hr

Z. Kapović
e-mail: zkapovic@geof.hr

A. Abdul Rahman et al. (eds.), *Developments in Multidimensional Spatial Data Models*, Lecture Notes in Geoinformation and Cartography,
DOI: 10.1007/978-3-642-36379-5_10,

1 Introduction

Changes in social and economic order in East Europe have caused great alterations, especially in the perception of real property ownership and real property transactions (Roić et al. 2008). Croatian land administration system faces a great crossroad because it demands for a single maintenance solution of digital cadastral plans converted from an analogue to a digital form a few years back. At the same time, Croatia is in the process of being admitted into full European Union membership (expected accession date is July 1, 2013).

The Land Administration Domain Model (LADM) is in the FDIS stage (FDIS = Final Draft International Standard): ISO FDIS 19152. This is the last stage before becoming an International Standard coming after a four-year standards development process within ISO/TC211 (Geographic Information—www.isotc211.org) and 6 years of preparation within the International Federation of Surveyors (FIG), while the original idea for such a standard was launched at the 2002 FIG congress in Washington D.C. This is why the Republic of Croatia also strives to keep track of the development of this international standard. Upon adoption of this standard, it needs to begin standardization of own land administration system in accordance with international norms.

Section 2 of this chapter describes historical and current situation of the Croatian land administration system. Section 3 gives brief information about LADM model and describes its general application while in the Sect. 4 discusses preparation of a Croatian national profile based on LADM.

2 The Croatian Land Administration System

Spatial data and data about real property in Croatia are managed in multiple registers with many end users. Croatian land administration system comprises of the cadastral and land book system. This section portrays most relevant registers of the spatial data and gives a short overview of the state of the data in these registers. There are also many land information systems in Croatia: LPIS which is quite agriculture related (Land Parcel Information System available at web site http://www.arkod.hr), Geoportal (available at web site http://geoportal.dgu.hr). National Spatial Data Infrastructure (NSDI) (available at web site: http://www.nipp.hr) is also currently being developed. NSDI is being developed in compliance with the directive 2007/2/EC of the European Parliament and of the European Union and of the Council of March 14, 2007 establishing an Infrastructure for Spatial Information in the European Community (INSPIRE).

Interested users of the Croatian land administration system are shown in Fig. 1. Their basic interests or responsibilities towards the Croatian land administration system are also portrayed.

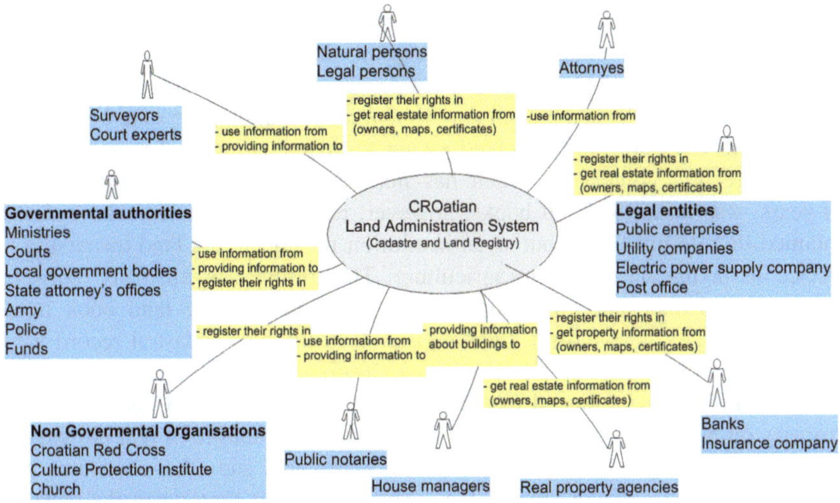

Fig. 1 Stakeholder model of the Croatian land administration system

2.1 History of the Croatian Land Administration System

Croatian cadastral system is based on the Germanic model of cadastre primarily because of its Austro-Hungarian heritage. Croatian cadastral system is similar to those in Austria and Germany. Development of land data registration in Croatia is conditioned by different countries of which regions of Croatia were a part during the past. This is why land data is registered with different dynamics and according to different conditions depending on the social structures in these countries. Attempts to establish a land cadastre resulted in an unprofessionally made and hence short lived Josephine Cadastre. Formation of the land cadastre in Croatia region under Austria-Hungary began with proclamation of the Imperial Patent (Grundsteuerpatent) on December 23, 1817, ordering surveys, land classification, and preparation of the cadastral record in all lands of the Empire. This date marks the beginning of the Franciscan Cadastre. Basic principles of this cadastre remained in use over a 100 years. Present day land registers are founded in the period from 1880 to 1900 based on these data (Roić et al. 2005).

The Land Register Law, not much different than Austro-Hungarian Grundesteuerpatent, is passed in the Kingdom of Yugoslavia in 1929, after World War I. In fact, this law was a translation of the Imperial Patent, so the cadastre merely continued to serve a tax purpose. Advancement of technologies in cadastral surveying, and especially development of numerical methods, sparked in this time period, enactment of a number of bylaws regulating these processes, some of which are used in practice even today.

After 1945 and the implementation of the radical changes in the social structure, attitude towards ownership and other real property rights also changed. The cadastre and land books did not enjoy support in this time period, but were instead neglected and finally terminated when private ownership was abolished and everything became communal. For this reason, the cadastre was not updated all until 1953 when Land Cadastre Legislation was passed. Lack of funds in the treasury caused this legislation since new sources of funding the state budget had to be found. Cadastral office revived owing to this situation, and revision of the cadastral record, content of which has not been maintained in the last circa 10 years, was initiated. Land books, however, remained neglected. The cadastre obtained its role in society, but as an institution for registering land ownership in service of taxing income from agriculture. This resulted in mismatch between cadastral land data and ownership data in land books, since land book items corresponding to the cadastral items registered in the new cadastral record were not updated from 1953 to 1991 (Roić et al. 2005).

Croatian Parliament passed the Law on State Surveying and Cadastre of Real Property on November 5th, 1999. This law defined the cadastre of real property as a register of land parcels, buildings and building parts, and other structures permanently on land or below the surface. This is the first time when the cadastre became a tool for registering real property as an object of law, and when it ceased to serve exclusively the tax purpose. Private ownership is in the foreground again as foundation for development of society, modern economy and overall progress. This law was in use for 8 years when law of the same title was passed in 2007, however, with some changes in terms of defining the real property cadastre, state surveys, and national infrastructure of spatial data. Numerous bylaws were passed alongside the law from 2007 regulating the geodetic reports, the Utility Cadastre, the Land Cadastre, the Real Property Cadastre, the administration of tasks regarding maintenance of state surveys and the real property cadastre, labels of country borders, execution of basic geodetic activities, etc. Alphanumerical part of the cadastral record for the whole Republic of Croatia is available on the internet (http://www.katastar.hr). Law from 2007 defines real property cadastre as a register of land parcels, buildings and other structures permanently on land or under surface, and of specific public rights and restrictions on land. Accordingly, one can notice that registration of building parts is omitted and public rights and restrictions are added as an entry in comparison to the law from 1999.

Cadastral plans originate from different time periods (Fig. 2). First employable plans, most of which are in scale 1:2880, arose with proclamation of the Imperial Patent and preparation of the land cadastre in Croatia region under Austria-Hungary. Seven coordinate systems with different starting points had to be used as surveys were performed without projection in plane rectangular system. Instructions for cadastral surveying (Katastral-Vermessungs-Instruktion) with appended legends and topographic keys (ZeichenErklaerung) defined basics of sketch representation of cadastral data in 1820. Few labels and some colours were used to mark building units. Changes and data updates of the cadastral plan according to the Cadastre Maintenance Law from 1883 were executed by crossing out old and colouring new condition in red, causing clutter and illegibility leading to systematic upgrade at the beginning of the 20th century in the Royal Lithographic Office in Vienna.

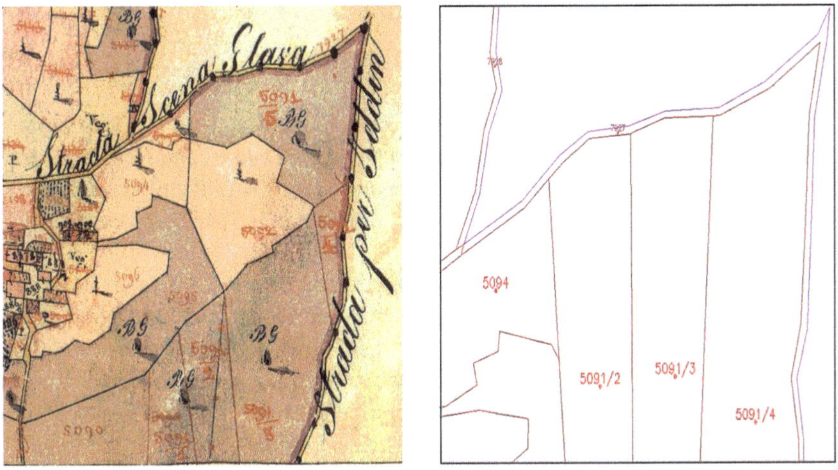

Fig. 2 Cadastral plans from different periods (*left* middle of 19th century, *right* present day—eletronic)

Kingdom of Yugoslavia introduced the Gauss-Krüger projection in 1924 with plans now prepared in scales 1:500, 1:1000, 1:2000, and 1:2500. Later, after World War II, technical purpose of the cadastral system during the Socialist Federal Republic of Yugoslavia demanded preparation of the topographical and cadastral plans and introduction of new real world features. By law of Cartographic Signs and Collection of Cartographic Signs defined content of these plans in 1976. Collection defined 346 labels classified in different categories defining sketches as dependent on scale. Poor physical state of cadastral plan pages and development of computer technologies stimulated preparation of scanned and vectorized digital plans in the nineties of the last century (Roić et al. 2005). Cadastral plans were digitalized for the whole Croatia by 2010.

Land books in present Croatia region have a long tradition. Imperial order was issued in 1850 initiating formation of land books in the Kingdoms of Hungary, Croatia, and Slavonia. Key legislations in this 160-year period are "Gruntovni red" (land book law) from December 15th, 1855, Law on Interior Design, Formation and Correction of Land Books from 1930, and Land Book Law from 1996. First attempts to establish land books were unsuccessful, but they were successfully established after land descriptions were transcribed from the cadastre. Land books are in present day available public registers of legal status of real property. Because of irregular data maintenance full confidence in accuracy of the registered data in land books is postponed till 2015 (Official Gazette of the Republic of Croatia, 1996). Purpose of land books is to enable simple, fast and safe legal transactions of rights registered on real property based on legal regulations. Land books are managed in specific departments of municipal courts. Land books for the whole region of the Republic of Croatia may be found on the internet address http://e-izvadak.pravosudje.hr/home.htm.

2.2 Current Situation of the Croatian Land Administration System

Pursuant to the Act on State Survey and Real Estate Cadastre (Official Gazette 116/99), the Decree on Defining the Official Geodetic Datums and Horizontal Map Projections of the Republic of Croatia was passed in 2004 (Officall Gazette 2004). The implementation of new official datums in practice is very complex and long-lasting process.

According to the Decree, the Croatian Terrestrial Reference System (HTRS96) was defined as a new positional system—Croatian realization of ETRS89 (European Terrestrial Reference System 1989—GRS80 (Geodetic Reference System 1980), ellipsoid heights), while the old Croatian coordinate system HDKS (Bessel, orthometric height) is still being used in practice until the complete transfer to the new reference system (Bačić et al. 2011). The new vertical system of the Republic of Croatia—HVRS71 (Croatian Vertical Reference System 1971) is determined by the geoid datum defined by the mean sea level for the epoch 1971.5 on five tide gauges equally distributed along the Adriatic coast (Dubrovnik, Split, Bakar, Rovinj and Kopar). Just as it is the case with the positional datum, the old vertical reference system defined by the tide gauge in Trieste as its starting point is still used (Table 1).

The cadastre registers in the alphanumerical part of the cadastral record data on parts of real property (apartments, common areas) should be identical to those in land books. Separate management and transcription of real property data often causes redundancy. Development of the Joint Information System of Land Books and Cadastre is on-going project in Croatia (Fig. 3) and it should reduce redundant data through single data storage of aforementioned registers. The most important purpose of this system should be, that once real property data is synchronized in the cadastre and land books, they no longer diverge.

Table 1 Coordinate systems officially in use in Croatia

Name of the coordinate system	Ellipsoid	Types of coordinates	Vertical datum
HTRS96 = ETRS89 = ETRF00(R05)	GRS80	φ, λ, h (X,Y,Z)	
HTRS96/TM	GRS80	E,N, H Transverse-Mercator projection	HRVS71
HDKS	Bessel	x, y, H Gauss—Krüger Projection	Trieste

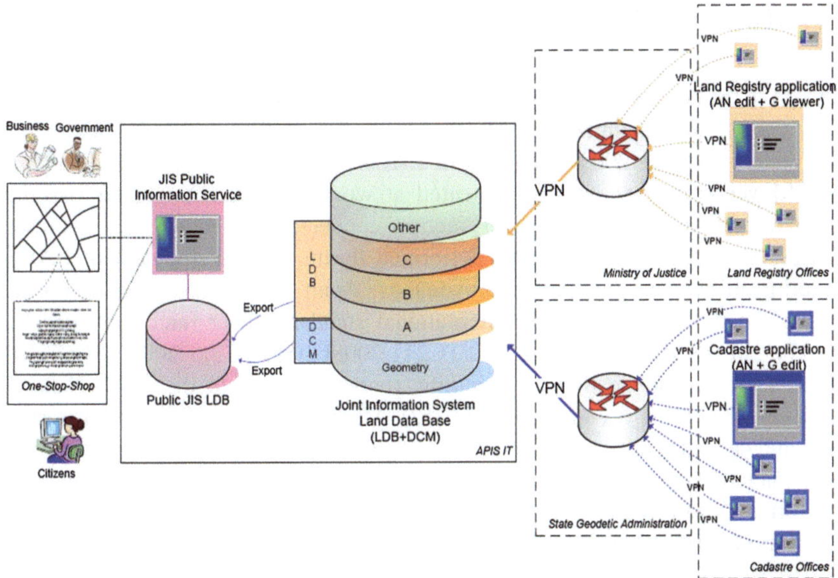

Fig. 3 Joint information system concept chart

3 LADM

Land Administration Domain Model (LADM) is a conceptual schema of land administration. LADM provides a reference model that will serve two goals: (1) to provide an extensible basis for the development and refinement of efficient and effective land administration systems, based on the Model Driven Architecture (MDA), and (2) to enable involved parties, both within a single country and between different countries, to communicate, based on the shared vocabulary (that is, ontology) implied by the model (ISO TC 211, 2011; Hespanha et al. 2008). With UML class diagrams relationship between people (LA_Party) and land (LA_BAUnit, LA_SpatialUnit) via rights, restrictions, and responsibilities (LA_RRR) are established (Fig. 4).

Fig. 4 Basic classes of LADM (ISO TC 211, 2011)

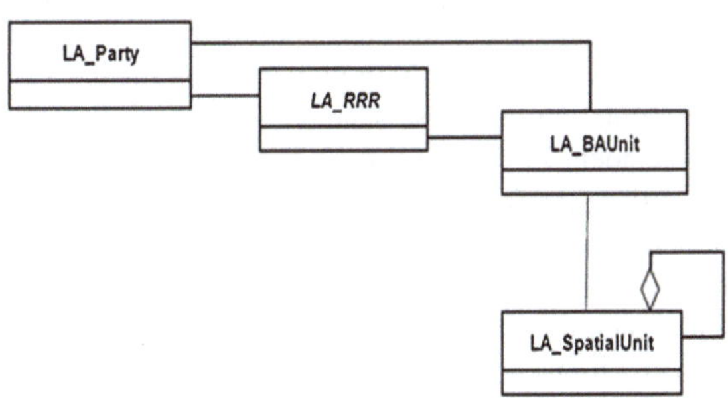

3.1 Short History of LADM

LADM is the conceptual schema of land administration which is derived from the Core Cadastral Domain Model (CCDM) (Thuy LE et al. 2012). It has undergone many changes since 2002 when idea of LADM (CCDM) was launched. The new LADM versions were usually presented at world geodetic congresses or other more important conferences and symposia connected to geodesy and geoinformatics. Many reviews have been performed during the development of LADM resulting in new insights and improvements. Apart from the versions published during the development of the international standard under (and published by) ISO/TC211 (ISO/TC211 2008a; ISO/TC211 2008b; ISO/TC211 2009; ISO TC211 2011) there are publications in scientific journals related to the LADM and its predecessor CCDM (Lemmen et al. 2012). Many entities, such as national mapping and cadastral agencies (NMCAs), research institutes and key industry players were included in development process of LADM. Therefore, LADM has been submitted to ISO/TC211 (Geographic information), for formal standardization and integration with other ISO/TC211 geo-information standards, such as ISO/IS 19107 Spatial Schema, ISO/IS 19108 Temporal Schema, ISO/IS 19111 Referencing by Coordinates, ISO/IS 19115 Metadata, and ISO/DIS 19156 Observations and Measurements (O&M).

3.2 Current Status of LADM

The Land Administration Domain Model (LADM) is close to become an International Standard. Currently it is published as a Final Draft International Standard: ISO FDIS 19152. More countries prepared their country profiles during the development of LADM.

3.3 Generating Country Profile Based on LADM

As a Platform Independent Model LADM provides generic view of the main objects associated with land administration. Therefore, it is independent of any considerations regarding implementation method (such as computer platform, data representation, database schema and other details). Being a generic and platform independent LADM has to be adapted to each country situation, but this country specific LADM will also remain platform independent model. Real implementation of such model will result in a Platform Specific Model based on defined hardware and software platform, while the process of implementation should be automated and computer aided with Computer Aided Software Engineering (CASE) tools and features from MDA approach.

When deriving a country profile from LADM (which often offers a number of possible choices) the following requirements should be taken into account (Hespanha et al. 2008):

- Existing technologies and extent of cadastral coverage in the country, mainly dealing with surveying and mapping regulations
- Institutional settings contributing to the definition of the set of Register Objects
- Legal and regulatory framework governing Rights, Restrictions and Responsibilities in Land, including both the fields of private and public law.

In short, when generating country profile based on LADM it is necessary to review administrative, legal and technical regulations of a particular country's existing land administration framework. In the following section the preparation of a Croatian country profile will be discussed.

4 Country Profile Croatia

This country profile is based on LADM and adds only some new classes, new attributes and new types to the code list (to differentiate LADM in Croatia from other models, their names are given with CRO_ as prefix).

4.1 Preparation of the National Profile

LADM is considered a Platform Independent Model and it is not meant to define a worldwide applicable standard for the Land Information System. It must be adapted to a specific country situation. A cadastral system like the Croatian one stands before great challenges with respect to implementation of the new geodetic datum and cartographic projection and establishment of the joint cadastral and land book information system. Croatia is herewith required to participate in land administration related international scientific and professional trends. This is why a country profile was to be defined (Fig. 5a, b).

4.1.1 Party Package

The main class of the Party Package is the basic class LA_Party (with party as an instance). LA_Party has a specialization: LA_GroupParty (with group party as an instance). Between LA_Party and LA_GroupParty there is an optional association class: LA_PartyMember (ISO TC 211, 2011). In the Republic of Croatia a party package presents owners, possessors and unregistered owners. The content of the party package is shown in Fig. 6. Party packages are inherited from LADM. A code list of party role types has some new types specific for the Republic of Croatia (Fig. 6).

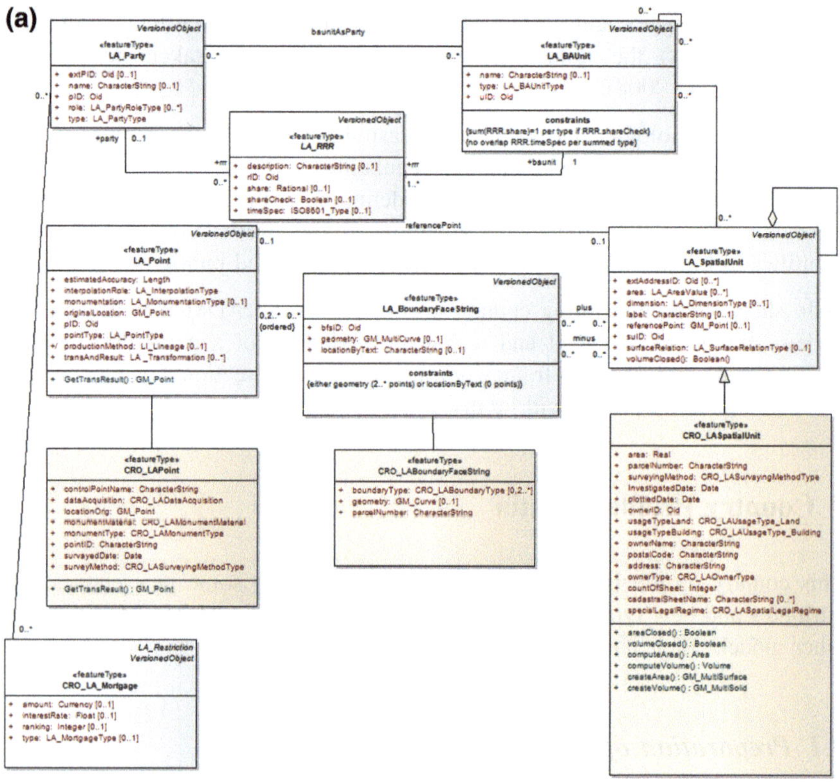

Fig. 5 **a** Country profile Croatia; **b** Country profile Croatia (code lists)

4.1.2 Administrative Package

The main classes of the Administrative Package are basic classes LA_RRR and LA_BAUnit. LA_RRR has three classes as specializations:

1. LA_Right, with rights as instances. Rights are primarily in the domain of private or customary law. Ownership rights are based on Law on ownership and other real rights.
2. LA_Restriction, with restrictions as instances. Restrictions usually "run with the land", meaning that they remain valid, even when the right to the land is transferred after the right was created. A mortgage, an instance of class LA_Mortgage, is a special restriction of the ownership right. It concerns the conveyance of a property by a debtor to a creditor, as a security for a financial loan, with the condition that the property is returned, when the loan is paid off.
3. LA_Responsibility, with responsibilities as instances.

Instances of class LA_BAUnit are basic administrative units. Baunits are needed, among other things, to register 'basic property units', which consist of

(b)

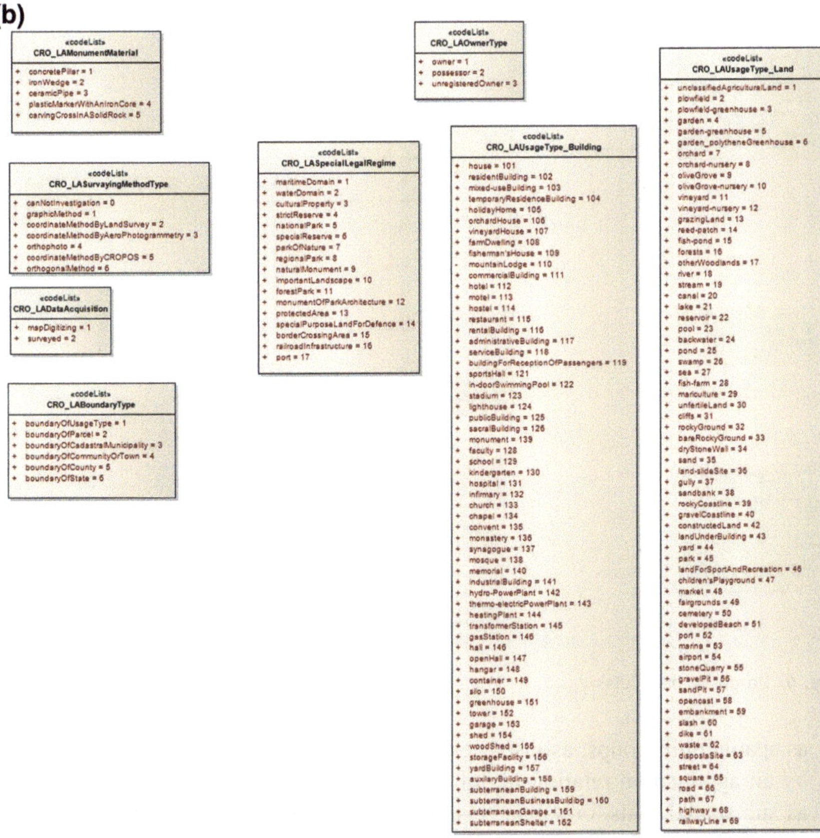

Fig. 5 (continued)

several spatial units, belonging to a party, under the same right. There must be a unique right for each baunit in order to establish a unique combination between an instance of LA_Party, an instance of a subclass of LA_RRR, and an instance of LA_BAUnit (ISO TC 211, 2011).

The attributes and relationships in Croatian administrative package are similar to LADM (Fig. 7). The administrative package contains rights, responsibilities, restrictions and administrative characteristics in land use.

4.1.3 Spatial Unit Package

The main class of the Spatial Unit Package is basic class LA_SpatialUnit, with spatial units as instances. LA_Parcel is an alias for LA_SpatialUnit. Spatial units may be grouped in two forms:

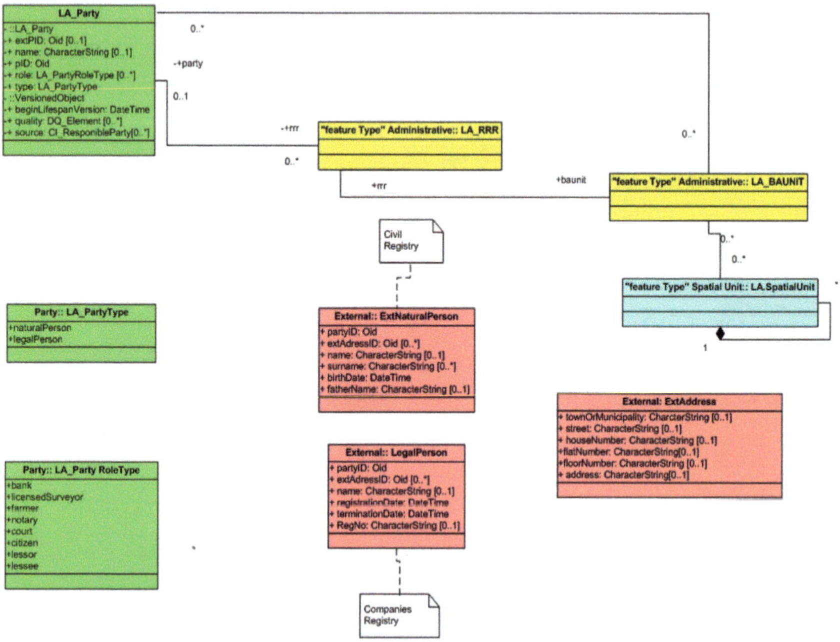

Fig. 6 Proposed party classes

1. as spatial unit groups, as instances of class LA_SpatialUnitGroup, and realized by an aggregation relationship of LA_SpatialUnitGroup onto itself.
2. as sub spatial units, or subparcels, that is grouping of spatial unit into its parts (ISO TC 211, 2011).

The spatial unit package presents the attributes of the parcels (Fig. 8), buildings and other structures associated with land. According to the Croatian laws some attributes, such as land use type, should be added. Basic spatial unit of a real property cadastre is a cadastral parcel. One cadastral parcel is a unit of a cadastral municipality or cadastral region at sea determined by a parcel number and its boundaries. Unique identifier of the cadastral parcel consists of an identification number of the cadastral municipality or cadastral region at sea and the parcel number. Boundaries of the cadastral parcel may be borders or other boundaries defined by legal relations on land surface (Official Gazette of the Republic of Croatia 2007).

4.2 Compatibility Testing

An abstract test suite complies with ISO 19105. Actual use of the LADM requires that an application schema, such as a country profile, is developed. The Annex A

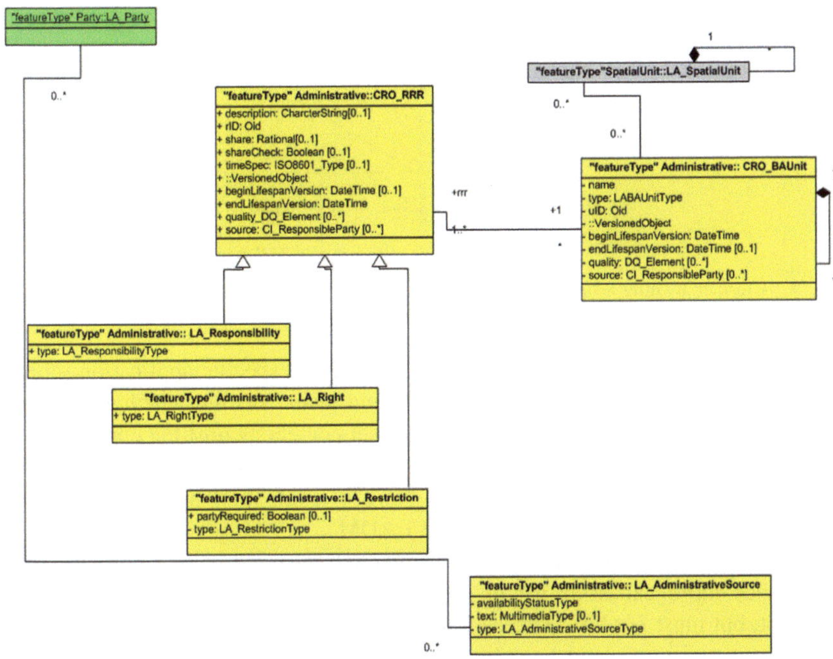

Fig. 7 Proposed administrative classes

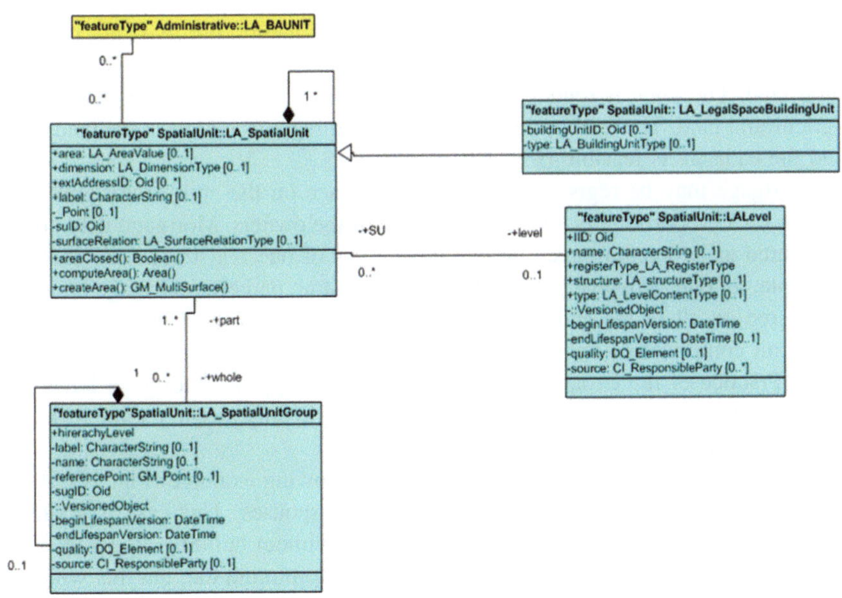

Fig. 8 Proposed spatial unit classes

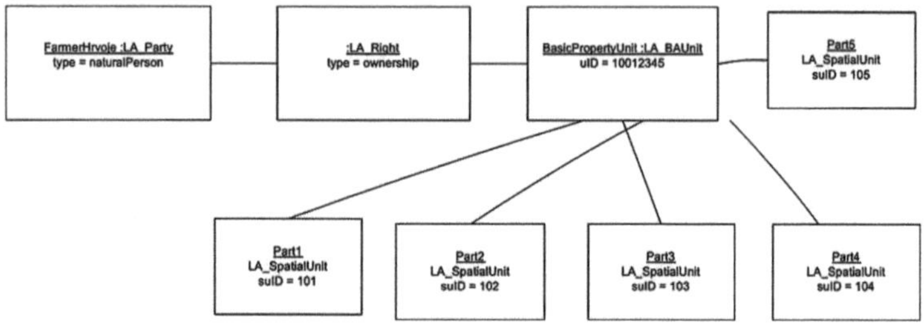

Fig. 9 A farmer owning a basic property unit with several spatial units

to the drafted standard specifies how to test whether a specific application schema complies with LADM in terms of package and level. Testing whether a specific data set complies includes checking the data set content against the corresponding compliant LADM application schema (Lemmen et al. 2012).

Further below are shown specific diagrams of compatibility tests between the Croatian land administration system and LADM.

An example from a land book—a farmer is registered in one property sheet which has more cadastral parcels. In nature, these cadastral parcels may be adjacent, but must not be. More cadastral parcels registered in one property sheet comprise a registered land unit. Moreover, the mentioned cadastral parcels from one registered land unit are registered in the cadastre in one possession sheet (Fig. 9). As the Republic of Croatia has a dual real property registry system (cadastre and land book) we have a similar situation in the cadastre—i.e. the farmer is registered as the owner and user of several cadastral parcels in one possession sheet. These parcels may be adjacent in nature, but must not be.

An example from a land book and the cadastre—leased state-owned agricultural land. The lease is registered based on the lease contract of the state-owned agricultural land. The testing showed that an unambiguous record in the land book and the cadastre is feasible (Fig. 10).

Mortgage may be registered as an encumbrance on the entire registered land unit or on a proportionally shared part of one of the owners. Mortgage may not be registered as an encumbrance on particular parts of one registered land unit (for example, if one property sheet has more cadastral parcels, mortgage must be registered on all of the cadastral parcels). Next figure shows that this registered land unit comprises of only one cadastral parcel, which is encumbered by mortgage. Practice of the Croatian banking system is to approve mortgage loans primarily in foreign currency (Euro, Swiss Franc), while only recently some loans are approved in local currency (Fig. 11).

Marriage relationships are presented in the following example (Fig. 12). Asset acquired in marriage is the common property of spouses. This case is a simple example of joint ownership over a simple parcel without building on parcel. This model could represent as well property over a condominium flat, but that will be a more complex example.

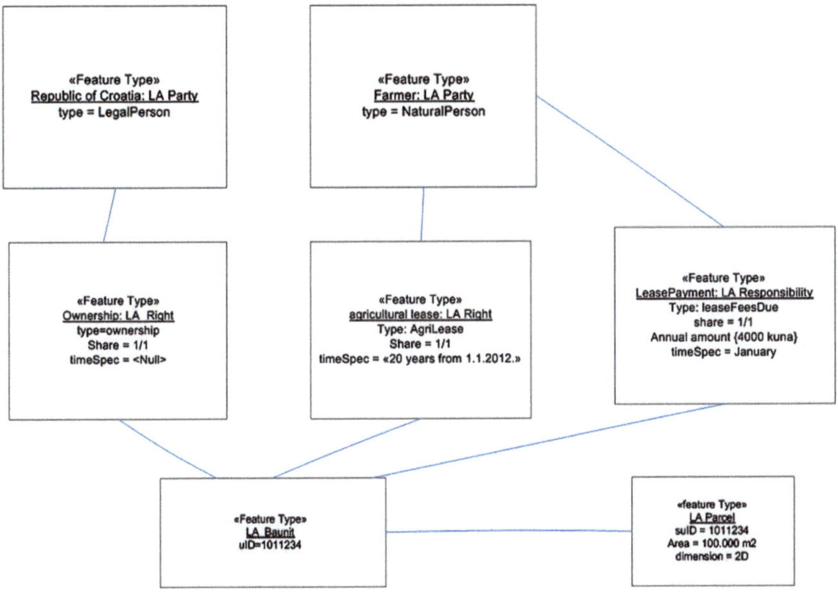

Fig. 10 Lease of a parcel owned by the Republic to the farmer

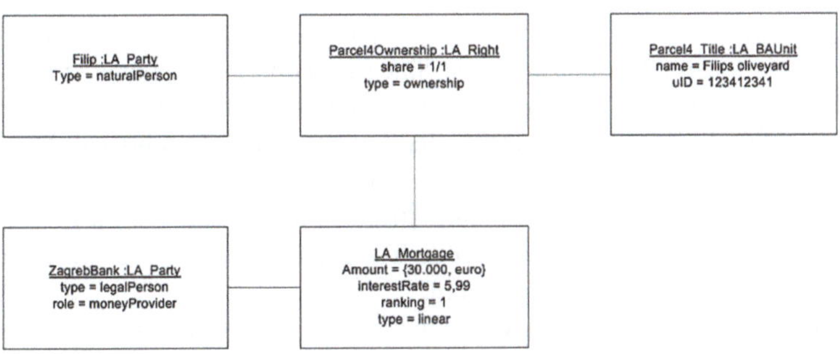

Fig. 11 Mortgage on parcel, bank included as party

During the whole testing process the following facts must be determined:

- What are the similarities and differences of the goals between the existing Croatian land administration system and LADM?
- What are user requirements for the Croatian land administration system?
- What LADM classes can be applied directly in the Croatian land administration system?
- What LADM classes have to be modified to be suitable in Croatian land administration system?
- How to design a land information system for the Croatian land administration system complying with LADM?

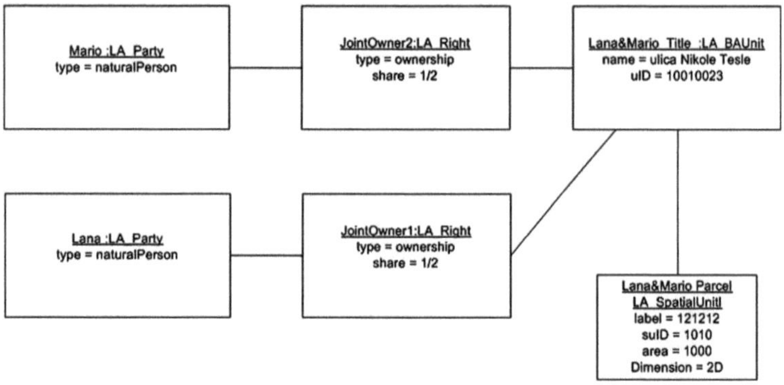

Fig. 12 Marriage relationships to property in Croatia

This chapter is initial research and gives the complete answer only to the first question. The main goal of the existing systems relevant to the Croatian land administration system is their compliance with the LADM as much as possible so that they can support social and economic development of Croatia. Achieving this goal implies consistency with LADM which is apparent from similarity of the objects and their relations in these systems. Just as in LADM Croatian land administration system includes connections between spatial unit (land parcels and other properties associated with land) and party (land users and managers) via rights, restrictions and responsibilities. Since LADM is conceptual schema there are differences between real-world situation and depending on the characteristics of the Croatian land administration system some of the elements can be applied directly while others have to be modified. Therefore, core elements of the LADM represented with packages, classes and attributes are extended with new classes, new set of attributes and types in the code lists according to the current situation of the Croatian land administration system.

Further research will give answer to the other questions and hopefully improve compatibility of the existing systems and LADM.

5 Conclusion

There are several countries that apply LADM to establish a country profile for their land administration system. This is initial research investigating compliance of the Croatian land administration system to LADM. Comparison between LADM and the Croatian land administration system established that there are similarities that may be utilized to further develop the Croatian land administration system. These results contribute to a more complete LADM based land administration model in Croatia and will contribute to the modernization process of the land administration system. In conclusion, LADM is compatible with the Croatian cadastre system, hence research on LADM implementation to the Croatian land administration system will be continued.

References

Bačić Ž, Marjanović M, Bosiljevac M (2011) Usage and upgrade of the Croatian position system. In: FIG working week, Marrakesh

Hespanha J, Van Bennekom J, Van Oosterom P, and Lemmen C (2008) The model driven architecture approach applied to the land administration domain model version 1.1—with focus on constraints specified in the object constraint language. In: FIG working week, Stockholm

ISO TC 211 (2011) ISO/DIS 19152, Draft international standard (DIS), geographic information—land administration domain model (LADM), ISO/TC 211 SN, 2011-01-20, pp 110

Lemmen C, Uitermark H, Van Oosterom P (2012) The final steps towards an international standard for land administration. In: FIG working week, Rome

Official Gazette of the Republic of Croatia (1996) Law on ownership and other real rights, No. 91

Official Gazette of the Republic of Croatia (2004) Decree on establishing new official geodetic datum and map projections of the Republic of Croatia, No. 110 and 117

Official Gazette of the Republic of Croatia (2007) Law on state survey and the real property cadastre, No. 16

Roić M, Tomić H, Mađer M (2005) Pregled katastarskih podataka. In: Third Croatian congress on cadastre, Croatian Geodetic Society, Zagreb, 421–427

Roić M, Fjalestad J. B and Steiwer F (2008) Regional Cadastral study. State geodetic administration, Zagreb, Croatia

Thuy LE P, Zevenbergen J, Lemmen C, Uitermark H, Quoc Binh T (2012) Investigating the conformity between the land administration domain model and the Vietnamese land administration system, FIG working week, Rome

GIS Application and Geodatabase for Archaeological Site Documentation System: Bujang Valley, Malaysia

Erna Warnita Bachad, Zulkepli Majid, Halim Setan, Albert K. Chong and Nurul Shahida Sulaiman

Abstract Recently, Geographic Information System (GIS) technology had becoming a common tool in various fields. The central and local authorities responsible for cultural heritage also embarked on creating an integrated information system, having GIS as one of the main infrastructure component. The objectives of the research are to capture the archaeological site (monument) spatial data using the technique of terrestrial laser scanning (TLS), as well as developing a 2D map and 3D visualization of Site Number 23. This research also examines the design and development of a GIS database and documentation system of an archaeological site. If all the data are made available in GIS, it can be combined and further analyzed by archaeologist as well as by the museum. The technique for spatial data capturing was TLS because its advantages to measure the distances and angles of objects without any contact and 3D visualization can be made from the point cloud data. A user-friendly interface of the system was completed by using Microsoft Visual Basic software. Thus, in this study, the archival documentation of an archaeological site which is Site Number 23 at Bujang Valley, Kedah was compiled into a GIS database.

Keywords GIS database · Archaeological documentation · Terrestrial laser scanning

E. W. Bachad (✉) · Z. Majid · H. Setan · N. S. Sulaiman
Faculty of Geoinformation and Real Estate,
Universiti Teknologi Malaysia, Johor Bahru, Malaysia
e-mail: ernawarnita@gmail.com

A. K. Chong
University of Southern Queensland, Toowoomba, Australia

A. Abdul Rahman et al. (eds.), *Developments in Multidimensional Spatial Data Models*, Lecture Notes in Geoinformation and Cartography,
DOI: 10.1007/978-3-642-36379-5_11,

1 Concept of Archaeological Documentation

Archaeology is a study of the past, primarily through the relics and it is also an array of methods and procedures that reflects the data which was obtained in order to study the evidence of the past (Hole et al. 1990). Archaeologists have been quick in realizing the great potential of high-resolution imagery for the detection and documentation of archaeological sites and features. A growing number of case studies with interesting results have evolved from ongoing archaeological field projects in recent years (Laet et al. 2009).

Meanwhile, archaeological documentation is a system that saves all the data and information about the archaeological site which may be in digital (softcopy) or hardcopy format. According to the Secretary of the Interior's Standards for Archeological Documentation, archeological documentation is a series of procedures applied to properties of archeological interest that involves planning, identification, evaluation or treatment. The documentation of a cultural heritage including the archaeological site may also be defined as the action of acquiring, processing, presenting and recording the necessary data for the determination of the position and the actual existing form, shape and size of a monument at a particular given moment in time (Georgopoulos et al. 2004).

Grabczewski et al (2002) argued that archaeological excavation is one of the method by which archaeologists answer questions about human history. Many techniques are available to archaeologists in determining the sequence of cultural developments over time, and one of them being used in Malaysia is stratigraphic excavation. Stratigraphic excavation relies on the elementary fact that in a series of earth layers or strata laid down on a given spot over a period of time, the sequential order of the strata from bottom to top of the deposit shows the relative age of any objects contained in the various layers (Laet et al. 2009). Current state of archaeological documentation in the historical-rich countries such as Rome, Greece and Peru are more advance compared to Malaysia. They have long been using the technique of GIS and digital system in documenting all the information about the archeological site. Thus, this research is aiming to develop a documentation system which focuses on database design and development of an archaeological site based on GIS and uses the data captured from the method of TLS. To achieve this aim, there are several objectives that have been completed are:

1. To collect the archaeological site (Site Number 23) data using TLS.
2. To develop a 2D map and 3D visualization of Site Number 23.
3. To design and develop a GIS database and documentation system of an archaeological site.

2 Advanced Technologies and GIS Techniques in Archaeology

The advantages of opening up the world of archaeology could be even greater than we can imagine. Instead of preserving the cultural heritage, archaeology also has the potential to expand knowledge from a previously unexplored point of view. Manuella Kadar (2008) stated that archaeology can be a great teacher if modern society has the awareness to learn the important of lessons from the past. History and archaeology can help us see why societies survive, change and collapse. In preserving ancient culture, there are numerous current techniques and technologies that have been used in capturing the spatial data of the archaeological sites. Some of the most significant techniques are photogrammetry, laser scanning, remote sensing as well as GIS. Photogrammetry and laser scanning are the solution used in architectural documentation in archaeological sites. According to Ergincan et al. (2010), new documentation technologies have been evaluated and analyzed by researchers for the application of architectural and archaeological projects. Warden and Woodcock (2005) stated that one of the major contributors to architecture education through the ages is the documentation of presented environmental elements and explained the applications and advantages of digital photogrammetry in the documentation of historical environments. Meanwhile, Nedal Al-Hanbali (2010) used the integration of digital photogrammetry and GIS in solving the requirement of 3D models with texture mapping. A research was conducted by Sauerbier et al. (2007) to record and model the Late Intermediate Period site of Pinchango Alto, Palpa, Peru by using a combination of image and range data that were captured using a mini helicopter and a terrestrial laser scanner. Both were equipped with a camera, allowed a fast yet accurate recording of the site and its stone architecture. The authors also mentioned that the resulting products can serve as a basis for archaeological analysis.

For this study, TLS was used as a tool to capture the spatial data. In recent years, the expansion of TLS technology is often seen as the solution to all cultural heritage measurement problems mainly because of the speed and accuracy of the method. Either as an individual technique or in combination with other measurement methods, TLS is extremely effective in recording complex and inaccessible objects. The method has also been successfully used in the protection and reconstruction of damaged sites, for example in UNESCO World Heritage Sites level such as the Bam Citadel, Iran (Behan et al. 2006). Moreover, the output which is point cloud data may be excessive for many applications.

One of the most important tools in archaeological and cultural heritage management is the application of GIS, in documenting as well as the analysis process and monitoring archaeological founds. The first comprehensive use of the GIS system in European archaeology was the work of Gaffney and Stančič (1991). Since then, GIS applications have become a widespread practice. Within archaeology, GIS can be deployed on three levels as discussed by Maschner (1996):

1. As a heuristic tool, used to receive answers to posed research questions (academic archaeology).
2. Within heritage offices (making data management more effective).
3. To visualize and popularize effects of inquiry.

According to Tantillo (2007), documentation is really important in archaeological studies to produce knowledge. Documentation is composed of three components, which are 'archaeological', 'technological' and 'geographical'. The author also stated that GIS is a computer system that has the capability to capture, integrate, store, edit, analyze and display the geographically-referenced information by using a location reference system, such as longitude and latitude. GIS makes data management and manipulation much easier, effective usability and seize an impressive potential for analysis. Thus, their rapid expansion and operation in various areas is of no surprise. González-Tennant (2009) used geodatabase for archaeology, examined the archaeologists responsibility to digital data sets, specifically GIS data, by creating a living document for documenting the archaeological finds. The latest issue in archaeological documentation is web-based GIS. By using web-based GIS technique, the availability of cultural heritage information can lead to an improvement in its management and preservation and also for an exposure and participation of the community. According to Brovelli and Magni (2006), this can be considered as an improving and easily accessible tool for the study of archaeological remains and also for the documentation. Internet technology is a useful tool for gathering and manipulation of data. There are also various software application for web based GIS, either open source software, commercial or some of them may be required as plug ins.

Archaeology, cultural heritage and historical science can help us understand the different processes involved in the creation of a sustainable environment. Other branches of sciences and social sciences, like chemistry, geology and zoology are widely used in archaeology. When an object is discovered, an archaeologist can use chemistry to determine its age, while botany or zoology can provide information about the surroundings where it was found.

3 Pilot Project: Bujang Valley, Malaysia

Department of National Heritage Malaysia is responsible for all archaeological activities in Malaysia. This department is responsible for exploring the possible area of archaeological sites, conducting excavation work, registration and preservation of artefacts and restoration of the sites. However, the development of archaeological activities in Malaysia is moving slowly in response to the use of conventional method for documentation and management of archaeological finds.

3.1 Current Situation in Bujang Valley

Lots of shrines were found in Bujang Valley but they are not digitally documented. As the sophistication of technology is spreading throughout the world, Malaysia is still using filing system to document the historical findings by capturing images and taking hardcopy notes. Conventional method is often delaying the processes involving data management especially the process of retrieving, updating and editing. Hence, an effective effort must be carried out to rescue the valuable inheritance for future generation. By and large, archaeological activities in Malaysia are moving slowly due to the small number of expertise in this field, the use of conservative technique for documentation purpose and there is a lack of awareness among Malaysians on the importance of preserving our cultural heritage.

3.2 Study Site

Bujang Valley, a place located in northern Malaysia (Kuala Muda, Kedah) is one of the most remarkable archaeological site. Archaeologists have already found a Buddhist stupa dating back to the First Century AD at the Bujang Valley. The site was discovered by James Low who was a superintendent in Seberang Prai. The sprawling ruins of an ancient Hindu kingdom dating back to 300 AD. Over 50 shrines and hundreds of relics are on displayed in the nearby Bujang Valley

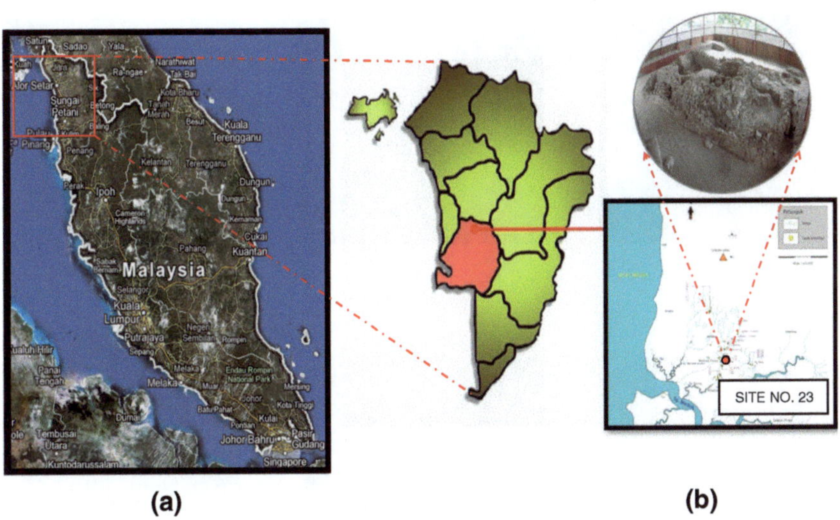

Fig. 1 Map of Peninsular Malaysia (**a**), map of Kedah and the distribution of the archaeological sites in Bujang Valley (**b**)

Archaeological Museum. Archaeological research indicates that an ancient Hindu-Buddhist kingdom ruled here as early as 300 AD. Relics of the kingdom found at the site and now on display at the Bujang Valley Archaeological Museum including inscribed stone caskets and tablets, metal tools and ornaments, ceramics, pottery, and Hindu icons (Jaafar 2003). Figure 1 shows the map of Kedah and the distribution of archaeological sites in Bujang Valley.

The selected archaeological site for this study is Site Number 23, located at Kampung Pengkalan Bujang. This site was discovered around 1980 but the excavation process was stopped at 1991. After several years, the excavation process continues again on 2004 until 2007 when it completely stopped. The types of artefacts found in this site are potteries, stoneware's, stones, qing pais, celadons, and some of metals and minerals.

The excavation process from 2004 to 2007 and also current state of the can be seen in Fig. 2.

The process conducted by the archaeologist including marking the site up to the documentation stage of Site Number 23 are listed below and in Fig. 3.

1. Measuring the site relevance to other features onsite.
2. Identifying the geographical-stratification.
3. Taking artefacts to laboratory for a cleaning up.
4. Process of cataloguing the artefacts.
5. Repairing the artefacts (if necessary).
6. Drawing, sketching and photographing artefacts.
7. Written documentation.

Fig. 2 The excavation process of site number 23(2004–2007)

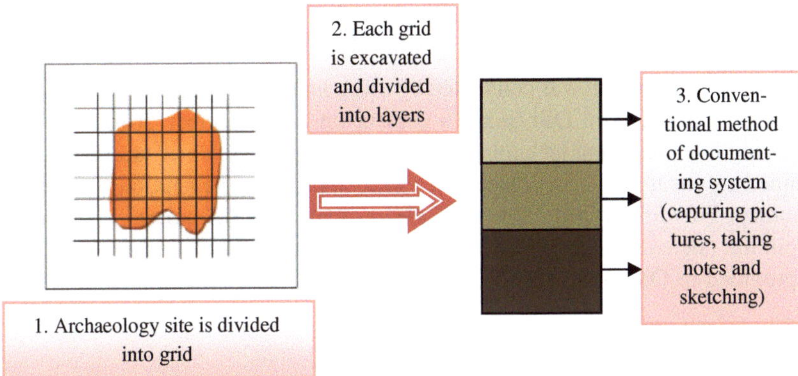

Fig. 3 Process conducted by archaeologist from marking the site up to the documentation stage

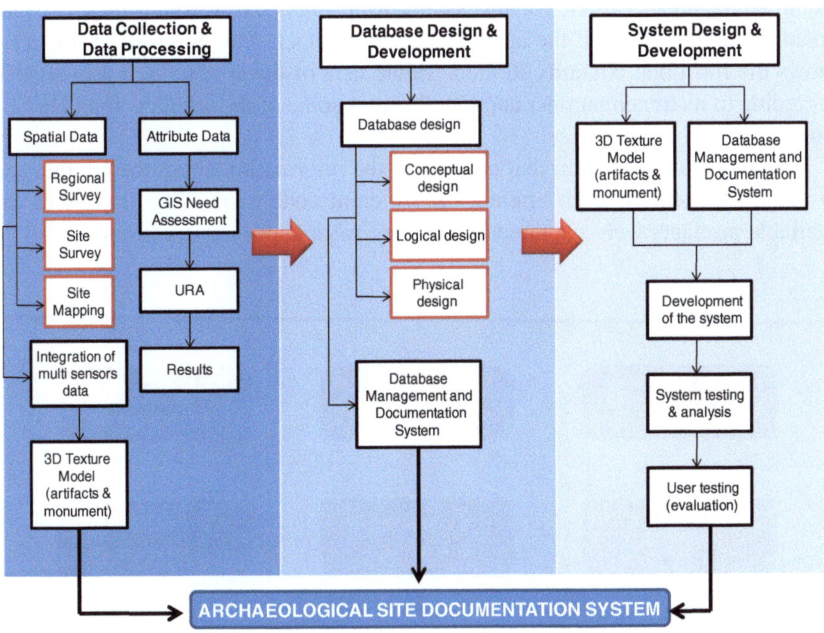

Fig. 4 Methodology

4 Methodology

This study is carried out to develop a documentation system of the archaeological sites, which comprises of the site (monument) and also the artefacts. An overview of the methodology can be summarized in Fig. 4.

4.1 Spatial Data Collection

Spatial data acquisition was conducted by using Faro Laser Scanner Photon 120/20 mounted with a Nikon DSLR 300 s camera on 21 December 2010. In projects involving laser scanning technology, the positions of the scanner must be carefully planned to ensure full coverage of the site, appropriate resolution and required accuracy. For this project, a total of eleven scans were acquired with scanning resolution one over four (1/4). The procedure of data capturing is shown in Fig. 5 as provided in Ong et al. (2010).

4.2 Attribute Data Collection

To develop the database, an appropriate attribute data collection procedure is required. As stated earlier, Bujang Valley Archaeological Museum used conventional method to store all the attribute data of artefacts Site Number 23. Figure 6 shows the form that contains all the attribute data of the artefact that was divided according to its trench number and Fig. 7 shows some of the artefacts found in Site Number 23.

Figure 6 shows the form that contain all the information about the artefact and to avoid redundancy, each artefact has different code name. The images of the available artefacts were captured before they were kept in a store room.

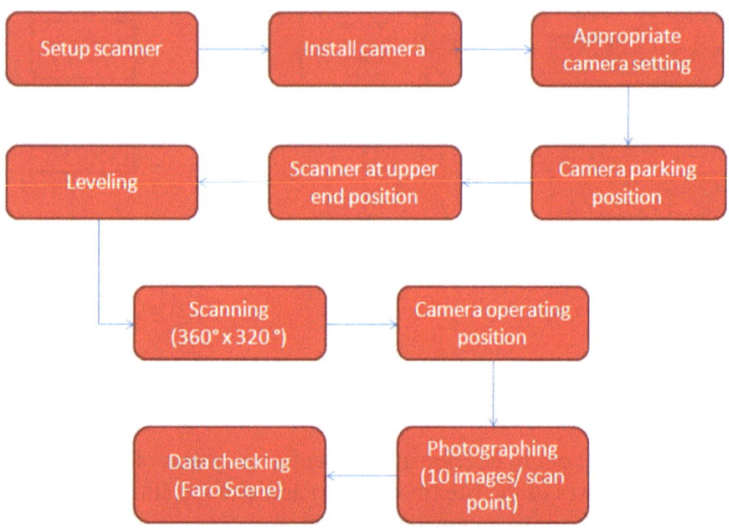

Fig. 5 The procedure of data capturing (Ong et al. 2010)

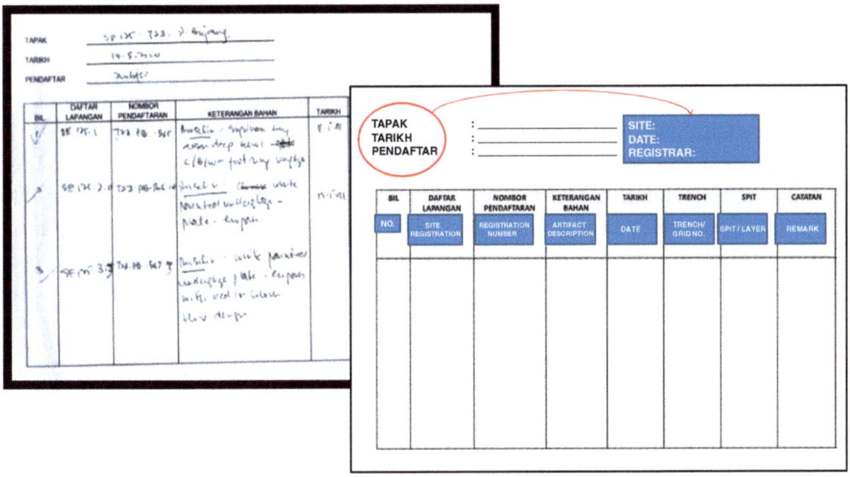

Fig. 6 Attribute data of artefacts site number 23

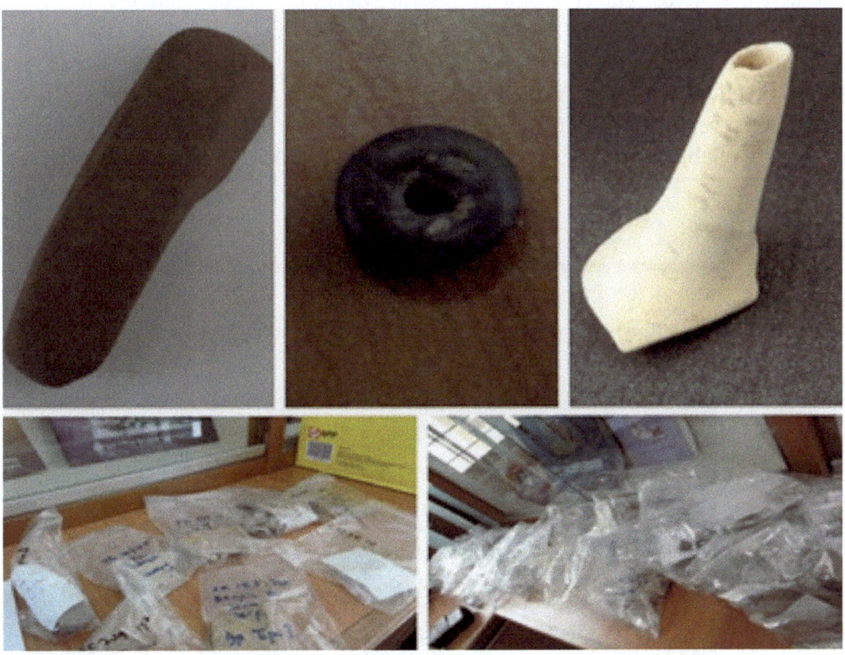

Fig. 7 Artefacts of site number 23

4.3 Data Processing

Faro Scene software was used in the registration and filtering process of archaeology sites data whereby the resulting scanned data is in.fls format. Registration is a basic process that needs to be carried out in order to merge the 3D point clouds of all the scanning stations. The filtering process involved basic checking of the data and filtering of unnecessary point cloud data such as backscatter and false return. This is because millions of measurement have been taken and must contain noise and errors. Figure 8 shows the registered and edited point cloud data of Site Number 23 in 3D.

4.4 Database Design and Development

4.4.1 User Requirement Analysis

As discussed by Maguire and Bevan (2002), understanding user requirements is the essence of information systems design and is important to the success of interactive systems. The core of a successful system is to understand the needs and requirements of the users. As specified in the ISO 13407 standard (ISO 1999), user-centred design begins with a thorough understanding of the needs and requirements of the users. Maguire and Bevan (2002) also stated that the process of URA is very significant due to the benefits such as improved productivity, enhanced quality of work, reductions in support and training costs and also

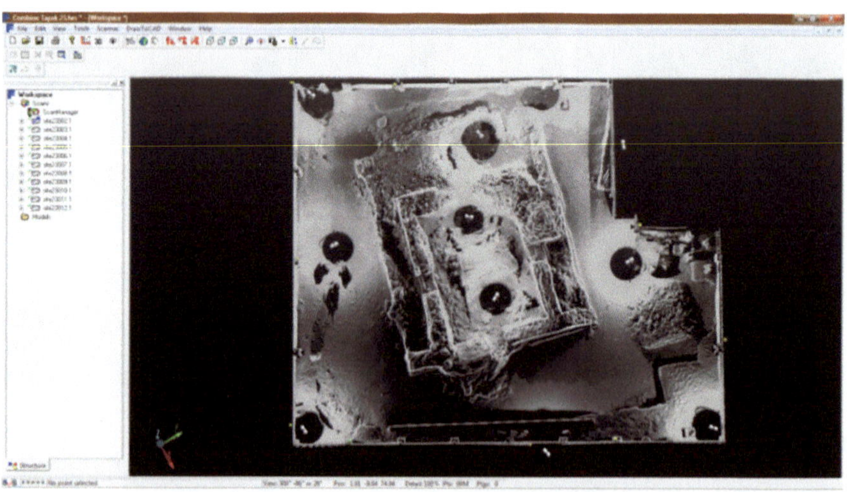

Fig. 8 3D point cloud model for site number 23

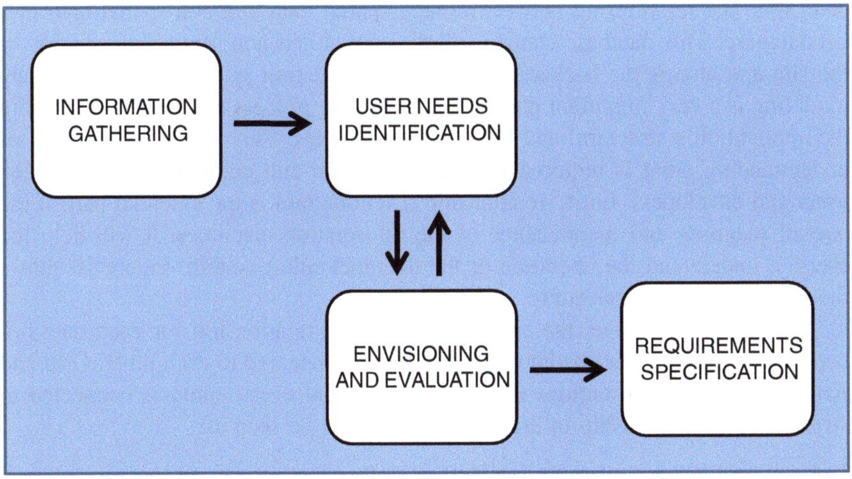

Fig. 9 General processes of URA

improved user satisfaction. There are four basic elements in the URA process (Fig. 9).

Information gathering began as soon as the study was conducted. The purpose is to gather background information of the users, who may be influenced by the system, which is the staff of Bujang Valley Archaeological Museum, as well as the archaeologist. User must be identified first so that the second element which is user needs identification can be classified. There are some approaches that can be used in identifying the user needs such as user surveys, questionnaire, interview and evaluation of existing system. For this study, an interview was carried out, and the interviewees were the representative of the museum who is the Director of Museum and Senior Museum Assistant. Besides that, a set of questionnaire was given to the Director of Museum, and it was answered personally by him. The questionnaire consists of:

1. Aim and objectives of the museum
2. Data and information available
3. Hardware, software and the existing system used
4. Staff background and skills in computer

The final stage is the evaluation of the results from the URA process. The hardware and software was also identified. The next step which is the database design and development need to be carried out.

4.4.2 Database Design

Database design is a process to produce a detailed structure of a database and includes conceptual design, logical design and physical design. It is important to determine the relationship between related spatial data and non-spatial data in a geodatabase. This database contains all the spatial and non spatial information of the site and also is the backbone of this documentation system. Conceptual data modeling is a very important phase to carry out the process of formally describing the application's structural and behavioral properties. This is for the purposes of understanding what is required for the application and communication between users and developers. Entity-relationship (ER) diagram is the essential part of the overall planning and maintenance of the information resources. It will help the users to understand the operation of the documentation system. Figure 10 shows the ER-Diagram for this study.

As seen in the above, the entity, attribute and relationship for each data has been determined, either spatial or non-spatial is connected to each other. Grid and Artefact are the main entities for this database and every entity is connected to Artefact. The relationship in this ER-Diagram can be seen as:

1. Each grid has one or more artefacts.
2. Each one artefact has its own information on excavation and registration.

According to Kadar (2008), physical database design starts from a given relational model which is the definition of a set of tables and their respective columns. The objective of physical database design is to fulfill the performance requirements of a set of applications by optimizing the use of the Database Management Systems. After the designing process, the next step is to input all the attribute data. The available image for the artefacts was also added to the database. Microsoft Office Access was used to key in the data. Then, Personal Geodatabase was developed in ArcCatalog.

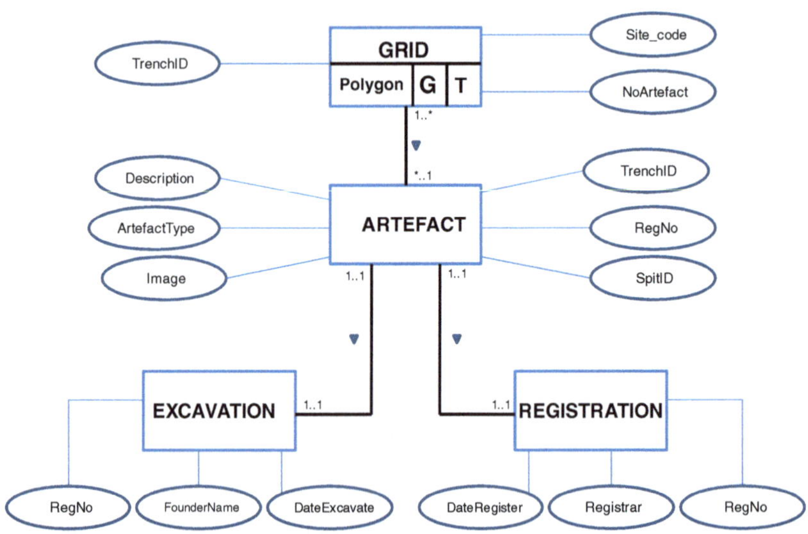

Fig. 10 ER-diagram

5 Results and Visualization

5.1 Archaeological Site Database

GIS can be used as a database management tool with a great flexibility. This technique have been applied with the best achievement to regional-scale archaeological survey, which is efficient for archaeological documentation as stated by Tantillo (2007). In this study, all the data are entered into a geodatabase with the data of GIS alphanumeric or textual and graphical or spatial information. It can manage, process and analyze such a complexity of archaeological record in space and time. Domains are used to set the possible inputs in the attribute tables of the feature class and relationship classes connecting one feature class attribute table to another as discusses in ER-Diagram.

The database consists of spatial and attributes data. This database will contribute many benefits to many organizations and individuals. Before this, user are depending on the conventional method, and if they want to retrieve, edit or update the database, they have to look for the artefacts and recording at the store rooms and obviously it takes more time. In addition, there are also problems in understanding the information because of the hand-writing technique. Now, the problem as such will not arise and this database can also link to other tabular information. Relationship that has been made in Arc Catalog will allow the query process as shown in Fig. 11.

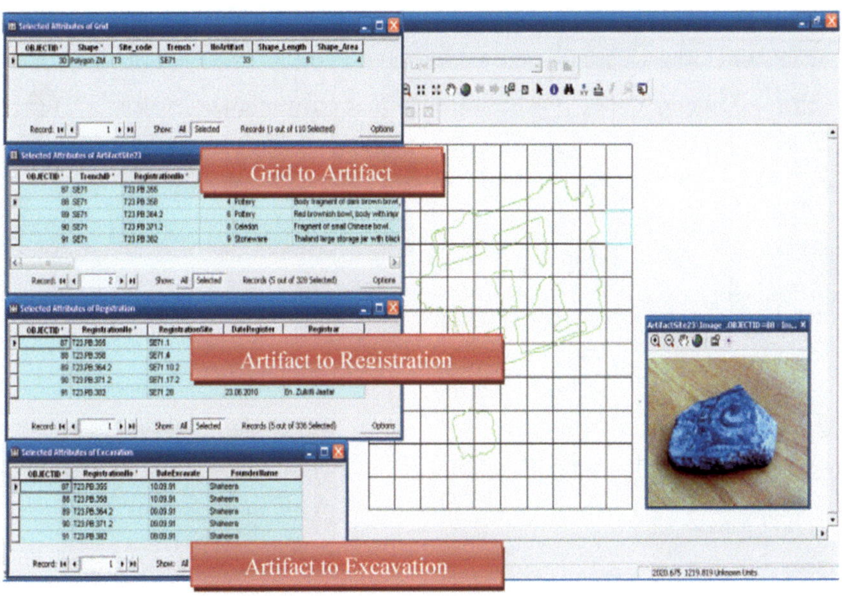

Fig. 11 The database and query process

5.2 Bujang Valley Archaeological Site Documentation System

The purpose of the user interface creation in system design and development is for the convenience of the user to access, manipulate and display the data effectively. There is a number of programming languages that can be used for creating user interface. The programming language that has been chosen to do the customization part is Microsoft Visual Basic 6.0 with embedded function of MapObject 2.0. To assist the user, a simple interface was created. In this study, there are several interfaces such as Login, Map Display, Video Player and Attribute's Display. User need to Login first before they can use this system. Figures 12, 13, 14 show some of the interfaces.

Login to the system is the first thing that a user has to do to use this system (Fig. 12). This is because the system was developed for the purpose of management and the data is confidential. Thus, the user has to key in the correct username and password before they can use it. Main menu will be displayed and user can see the information about the museum and also the video for 3D visualization of the site. Spatial Data Interface (Fig. 13) is the main interface of the system. In this interface, there is a map display that has the information of Site Number 23 (map and grid). The functions available are Identify, Zoom In, Zoom Out, Full Extend, Pan, Query and Print. All these functions were made to help user in finding the information and get to know the location for the spatial data. Apart from that, there

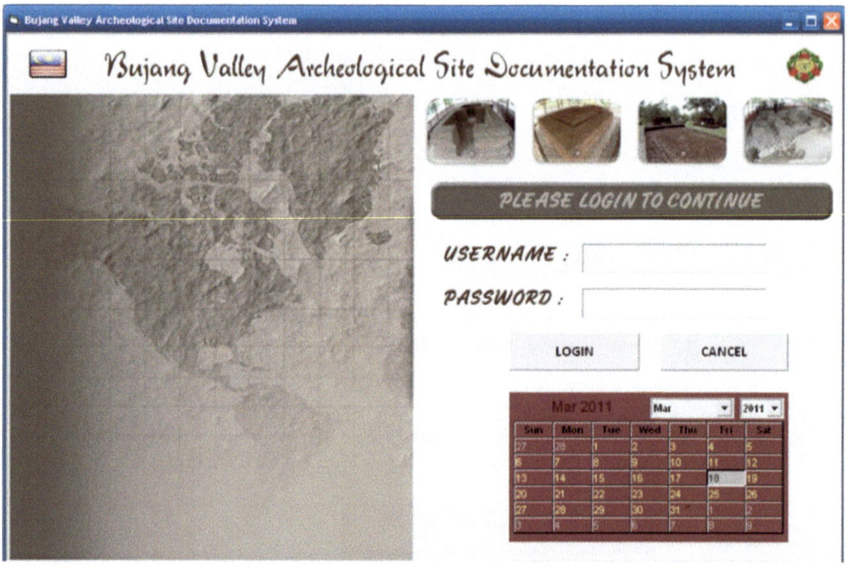

Fig. 12 Login interface

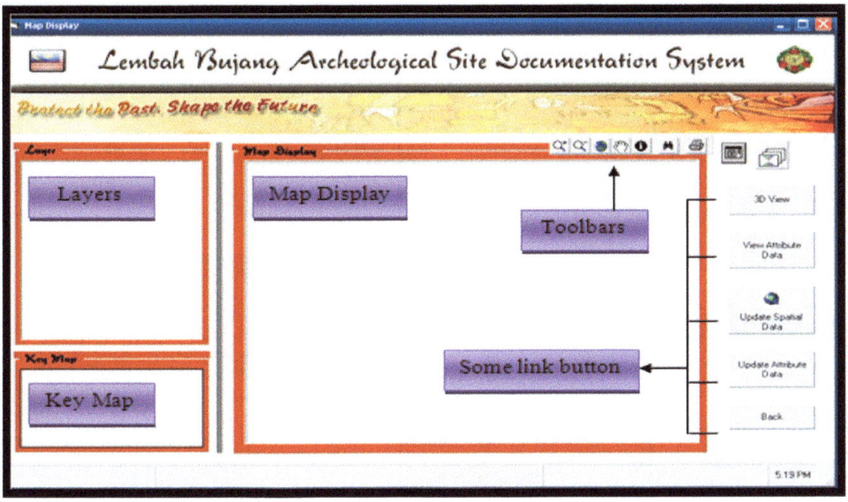

Fig. 13 Spatial data interface

are some buttons that will help user to access directly to the attribute data, 3D visualization and to update and edit the attribute data of the artefacts.

The map display used the programming that calls the shape file data to the map display form. The color of the layer and also the order of the layer in the map display interface were included. The toolbars has some functions such as Zoom In, Zoom Out, Full Extend, Pan, Identify, Query (Fig. 14 shows the query interface) and Print. Every symbol has its own embedded code to provide the functionality in handling map layers. Each function was programed by using the code that was prepared along with the MapObject component.

The Query function was made so that the user can make the query process faster and easier. This is because the map display should be more effective and interactive in delivering the information to the user. User can choose either to 'Query by Trench ID', which will be directly connected to the map display or 'Query by Type of Artefact' that will display the attribute data and image of the artefact.

Every system development has to be tested to determine its performance and functionality. There are a few steps to test the functionality and performance of the system being developed for Bujang Valley Archeological Museum namely query, identification, spatial select and data updating. In addition, the performance of the system for different processors has been conducted. In this last phase, some simple queries will be applied to test the integration between the systems with the attribute data in geodatabase. In terms of accuracy, this system shows that the results are correct as it follows the users command. This is according to the result from the system, for example in the process of 'Identify', 'Query' and also in data updating. It shows an accurate result, both in spatial and non spatial data.

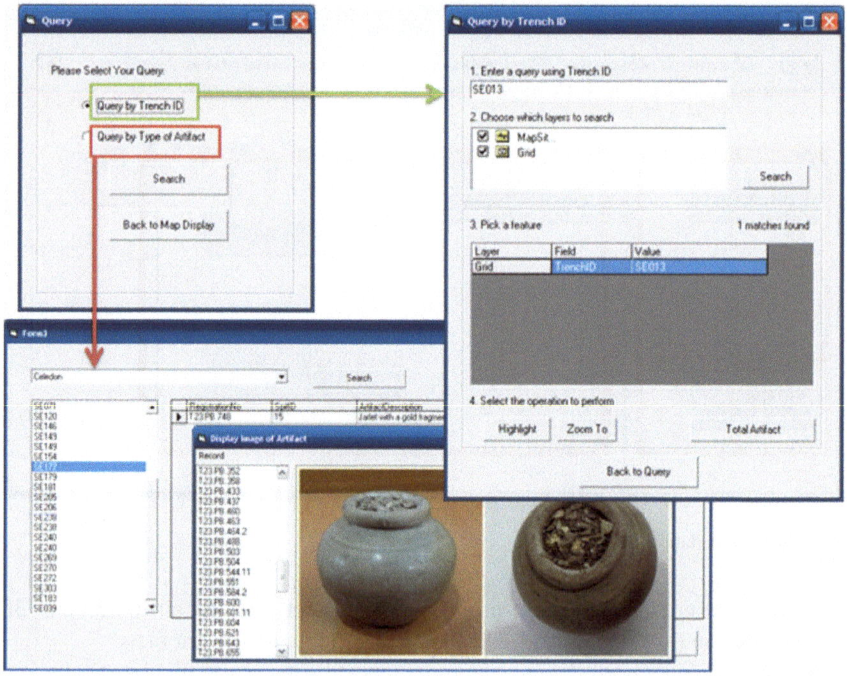

Fig. 14 Query interface

6 Conclusion

Documentation of the archaeological site is very important in order to protect and preserve, as archaeology and cultural heritage protection is an important issue around the world today. Developing an archaeological site documentation system and implementing TLS method for spatial data collection were the aim and objectives in this research. GIS is becoming a powerful and very effective tool for documentation and analysis. The outputs which are database, 3D visualization and documentation system developed for Site Number 23 can be broadly applied to other sites that present similarities in excavation methodology. The systematic exploration and representation of archaeological data in GIS environments can reveal a new opportunities for the organization in many aspects, especially in management and analysis. In this sense, the proposed workflow allows the recording of archaeological observation in a digital GIS environment, which can support tools for the exploration of archaeological information and provides the ability for further analysis.

Documentation should be considered as an integral part of a greater action in general documentation of the cultural heritage (Georgopoulos et al. 2004) as well as the archaeological site. There are some recommendations for future research such as enhancing the graphics of the interface, adding more applications for example being able to compare between two artefacts in 3D model and adding some spatial analysis process such as archaeological site prediction.

References

Al-Hanbali Nedal (2010) Texture mapping and implementation aspects for 3D GIS applications. Surveying and Geomatics Engineering Department, Al-Balqa Applied University, Jordan

Behan A, Moss R (2006) Close-range photogrammetric measurement and 3D modelling for Irish medieval architectural studies. In: The 7th international symposium on virtual reality, archaeology and cultural heritage VAST (2006)

Brovelli MA, Magni D (2006) An archaeological web GIS application based on mapserver and post GIS. In: The International Archives of the Photogrammetry, Remote Sensing and Spatial Information Sciences, Vol. XXXIV, Part 5/W12 25–27, Dresden, Germany, September 2006

Ergincan F, Çabuk A, Avdan U, Tün M (2010). Advanced technologies for archaeological documentation: patara case. Sci Res.Essays 5(18):2615–2629

Gaffney V, Stančič Z (1991) GIS approaches to regional analysis: a case study of the island of Hvar (2nd Edition). Oxbow Books, Park End Place

Georgopoulos A, Ioannidis G (2004) Photogrammetric and surveying methods for the geometric recording of archaeological monuments, archaeological surveys. The Olympic Spirit in Surveying, FIG working week 2004, Athens

González-Tennant E (2009) Using geodatabases to generate "living documents" for archaeology. A Case Study from the Otago Goldfields, New Zealand

Grabczewski E, Cosmas J, Santen PV, Green D, Itagaki T, Weimer F (2002) 3D MURALE: multimedia database system architecture. Assoc Comput Mach. ACM 1-58113-447-9/02/0009

Hole Frank, Heizer Robert (1990) Arkeologi prasejarah: satu pengenalan ringkas. Dewan Bahasa dan Pustaka, Kuala Lumpur

ISO (1999) ISO 13407: Human-centred design processes for interactive systems. International Standards Organisation

Jaafar Z (2010). Director of Bujang valley archeological museum

Kadar Manuella (2008) Data modeling and relational database design in arc haeology. Acta Univ Apulensis J 3(2008):73–80

Laet V, De and Lambers K (2009) Archaeological prospecting using high-resolution digital satellite imagery: recent advances and future prospects. A Session Held at the Computer Applications and Quantitative Methods in Archaeology (CAA) Conference, Williamsburg

Maguire M, Bevan N (2002) User requirements analysis: a review of supporting methods. In: proceedings of IFIP 17th world computer congress, Montreal, Canada, pp 133–148, 25–30 August 2002

Ong CW, Cheong SC, Zulkepli M, Halim S (2010). 3D documentation and preservation of historical monument using terrestrial laser scanning. Geoinf Sci J 10(1):73–90

Sauerbier Martin, Lambers Karsten, Eisenbeiss Henri, Kupferschmidt Denise, Gaisecker Thomas, Sotoodeh Soheil, Hanusch Thomas (2007) Combining photogrammetry and laser scanning for the recording and modelling of the late intermediate period site of pinchango alto. Palpa, Peru

Tantillo MD (2007) GIS application in archaeological site of solunto. XXI International CIPA Symposium, Athens

Warden R, Woodcock D (2005) Historic documentation: a model of project based learning for architectural education. Landscape Urban Planning. 73:110–119

Maschner HDG Geographic information systems in archaeology, New Methods, Old Problems, pp 1–21

A Review and Conceptual Framework for Generalization of Maps

Siddique Ullah Baig, Alias Abdul Rahman and Edward Eric Duncan

Abstract Generally, National Mapping Agencies (NMAs) produce and maintain topographic and cadastral maps. Regular update of map products is the most critical problem faced by NMAs. Therefore, implementation of new technology to automate map generalization has become crucial in updating smaller-scale maps from frequently updated larger-scale maps. Additionally, customized data products can be produced. Implementation of single database and derivation (generalization) of multiple-scale maps from one database can improve the efficiency and maintenance of derived data products. However, consensus between NMAs on a common generalization technique has not been achieved due to incompatibility and complexity of data models. Definitions of terms, technology and organizational structures of the NMAs across the globe are different. This chapter reviews the pertinent generalization efforts made by a number of European NMAs. A framework is proposed based on two correlated models: process-oriented generalization models and object-level-oriented generalization models, which is related to level of map objects. A portion of this framework is implemented and tested on datasets of *Kelantan* state in Malaysia to produce generalized maps. Generalization operations (e.g., simplification, aggregation, smoothing, etc.) are applied with the aim to generalize polygons and transportation network (e.g., roads, etc.). The proposed framework could be adopted, implemented and tested on datasets maintained by Malaysian National Mapping Agency known as JUPEM (*Portal Rasmi Jabatan Ukur Dan Pemetaan Malaysia*).

Keywords Generalization · Cartography · Scale · Simplification

S. U. Baig (✉) · A. Abdul Rahman · E. E. Duncan
3D GIS Research Lab, Faculty of Geoinformation and Real Estate,
Universiti Teknologi Malaysia, Johor Bahru, Malaysia
e-mail: siddiquebaig@gmail.com

A. Abdul Rahman
e-mail: alias@utm.my

E. E. Duncan
e-mail: edwduncan@yahoo.com

A. Abdul Rahman et al. (eds.), *Developments in Multidimensional Spatial Data Models*, Lecture Notes in Geoinformation and Cartography,
DOI: 10.1007/978-3-642-36379-5_12,

1 Introduction

Generalization is defined as a process which realizes transition between different models representing a portion of the real world at decreasing detail, while maximizing information content with respect to a given application (Weibel and Dutton 1999). In general terms, automatic generalization is the process of producing maps from one scale to another from a geodatabase within a digital environment. The goal, is to preserve the essential characteristics, behavior of objects, minimize existing constraints (preservation, legibility, and topological) and violations during scale transition. During the map generalization process, the number of objects is reduced, relevant information is highlighted and irrelevant information is suppressed based on the application, user and the generalization performed on demand and on-the-fly (Sester 2002).

A number of frameworks or strategies have been introduced to implement generalization process. McMaster and Shea (1992) highlighted three major issues in map generalization with regards to "'why', 'when' and 'how' maps should be generalized". The framework proposed by Brassel and Weibel (1988) addressed the last issue of 'how maps should be generalized'. This framework has been widely recognized as being feasible for computerization (Li 2007).

Figure 1 illustrates an example of the generalization process for approximating a curve by line segments through reduction of nodes similar to Li (2007). The original line segment (a), some of the nodes (red) of the line segment (b) are removed as part of generalization process and a generalized line segment (c) produced.

National Mapping Agencies (NMAs) are responsible for producing and maintaining topographic and cadastral data. Efforts are being made successfully at European NMAs to derive small scale maps from large scale maps within a digital environment. This increases the efficiency of map production at multiple scales and customized data products. Problems faced by NMAs are: (1) A common generalization technique cannot be applied in all NMAs due to different scales, different definitions of similar terms, and different terms for similar meanings, different technological and organizational structures of the NMAs across the globe (Stoter 2005). (2) However, this topic is mostly subject to research and only

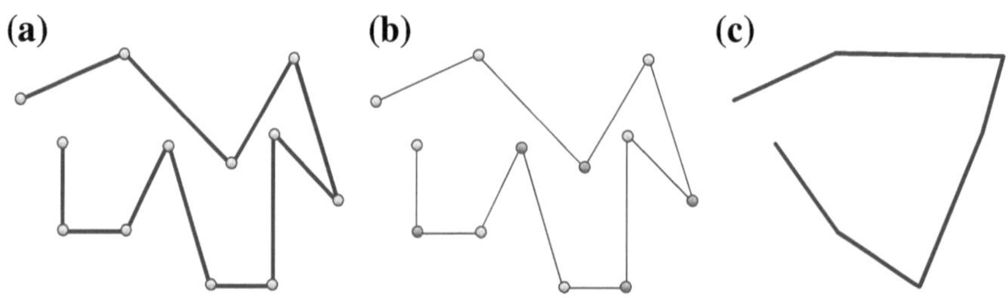

Fig. 1 Approximation of a curve by line segments through node reduction. **a** Original line. **b** Node reduction process. **c** After node reduction

specific research results have found their way into practice (Lecordix et al. 2007; Regnauld and Revell 2007). This is due to data complexity of completed datasets and incompatibility of data models within NMAs. Three important aspects of automated generalization are to update maps at smaller-scales by deriving information from frequently updated larger-scale maps (Li 2007). For this purpose, ideally only one detailed database should be updated frequently and maps at other scales are only updated on demand. Single scale database can be implemented with the help of automatic generalization and improve the maintenance of all data products (Foerster et al. 2010).

A framework based on a combination of both process-oriented generalization models and object-level-oriented generalization models for map and database generalization similar to Brassel and Weibel (1988) is proposed. It is composed of five processing steps: (1) Structure of database or map, which is controlled by target-scale and generalization rules as generalization objectives. (2) Essential features from structured spatial data are recognized based on conflicts, structure of target-database or maps. (3) Compilation of rules and procedures from an operation library. (4) Operation *execution,* consists of a sequence of operational steps compiled from generalization functions stored in the operations library. (5) *Data display,* transforms the target data into the target generalized map or database. A portion of this framework is implemented and tested on datasets of *Kelantan* state in Malaysia, to produce generalized maps. Simplification of polygons and transportation network, aggregation of simplified polygons and smoothing of polygons are a few generalization operations performed as part of this research. The data used for this purpose was a small area of *Kelantan* state, in Malaysia of size 6.5 by 3.5 km. Input for simplification process and aggregation operations was a set of 7,014 polygons with transportation networks such as roads and railway lines.

State-of-the-art—generalization of maps is discussed, followed by a detailed discussion about different generalization models and approaches adopted by the NMAs of Germany, France, Spain, and Denmark in Sect. 2. A framework for generalization of maps is proposed in Sect. 3 followed by implementation and results in Sect. 4. Finally, a summary is presented in Sect. 5.

2 State-of-the-Art: Generalization of Maps

During the map generalization process:

- Amount of objects are reduced
- Relevant information is highlighted
- Irrelevant information is suppressed based on the application, the user and the generalization is performed on demand and required (Sester 2002).

Perkal (1966) first, proposed the concept of objective generalization, it is difficult to establish the precise date on which research on digital map generalization started. Two very important publications appeared the same year: Tobler (1966)

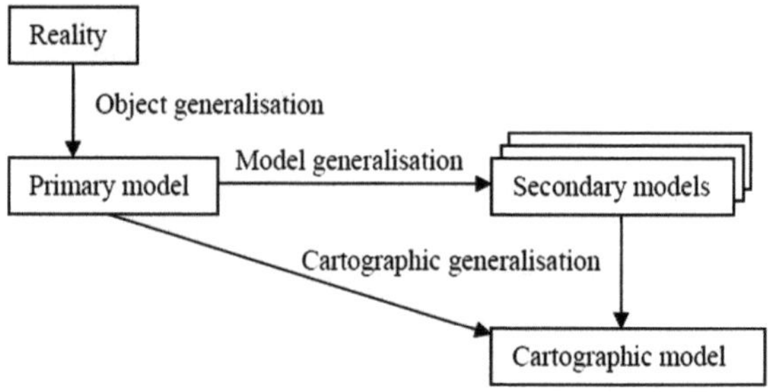

Fig. 2 Model and cartographic generalization of [adopted from Grünreich and Powitz (1992)]

which was explicitly for the smoothing of numerical data to generate the generalization effect; and Töpfer and Pillewizer (1966) for describing the relationship between map scales and number of symbols on map (Li 2007). An overview about the concepts related to generalization of cartographic maps is presented in (Mackaness et al. 2007) and (Weibel and Dutton 1999). There was only one generalization model before (Grünreich et al. 1992) developed a new model separating generalization into model and cartographic generalization (see Fig. 2). The model for the generalization deals with modification of geo-data and the cartographic model deals with the production of GIS maps with different scales, while avoiding cartographic conflicts.

Classification of generalization functions or operators such as simplification, enhancement, displacement and aggregation varies, based on the conceptual understanding of the cartographer. According to (Foerster et al. 2010), a common understanding of classification functions is not existent. Previously, before the advent of software tools for generalization, cartographers rely on manual generalization which could again generate different results from one cartographer to another. In a few NMAs e.g., Survey of Pakistan, generalization process is carried out manually for multiple-scale maps. Even, some NMAs such as The Netherlands (TDK) apply additional constraints manually if existing ones becomes incompatible for some areas within the map. Application of constraints did not cover what should be moved, deleted or enhanced keeping conflicts into account in manual generalization. Currently, interactive generalization using software such as ArcGIS, *Change/Push/Typify, Radius Clarity* and *expand* is carried out in this research.

2.1 Implementation of Generalization Processes at NMAs

It is hard to define a generic generalization process applicable to (m)any topographic data set(s) (Foerster and Stoter 2008). According to (Foerster et al. 2010; Lecordix et al. 2007) and (Regnauld and Revell 2007), generalization of cartographic maps, is subject to research at NMAs due to differences in the classification of generalization operators and incompatibility of data models, even within the same NMA Even, different results are produced for same generalization rules given to two cartographers due to interdependencies between different (types of) objects and the mutual effect of generalization rules. Therefore, (Stoter 2005) suggests, it is not easy to formalize generalization rules to produce the same outcomes. However, cartographic societies have concluded a number of generalization evaluation methods for a long time. Foerster et al. (2010) presented the challenges for dynamic generalization at European NMAs integrating a qualitative and quantitative analysis. Qualitative analysis was carried out about the trends and policies on dynamic generalization practiced by NMAs. Outcomes based on initial datasets, supported scales and aspects of geometry types were taken into consideration as part of quantitative evaluation. Operators that create generalized features and contextual operators that inherit a network—based structure are the main challenges for cartographic generalization.

The generalization tasks performed by cartographers are subjective; therefore, the way they are solved is different (Kilpeläinen 2000). State-of-the-art on generalization within NMA is difficult to answer due to their specific characteristics. However, all NMAs distinguish between database and cartographic generalization but the difference is ambiguous (Stoter 2005). A number of researchers have conducted surveys and research to know generalization practices at NMAs across the globe especially one conducted by [see Stoter (2005) and Kazemi et al. (2007)]. Figure 3 illustrates an overview of the different scale transitions applied at NMAs.

Generalization efforts made at a number of NMAs (e.g., Germany (BKG), France (IGN), Spain (ICC), and Denmark (KMS)) shows that automatic generalization can be achieved with the aim to generalize and updates dataset/databases without renewing production lines.

Common goals have been identified for all mapping agencies in Germany and this has led to specifications for a Base-DLM and a DLM50 for the whole of

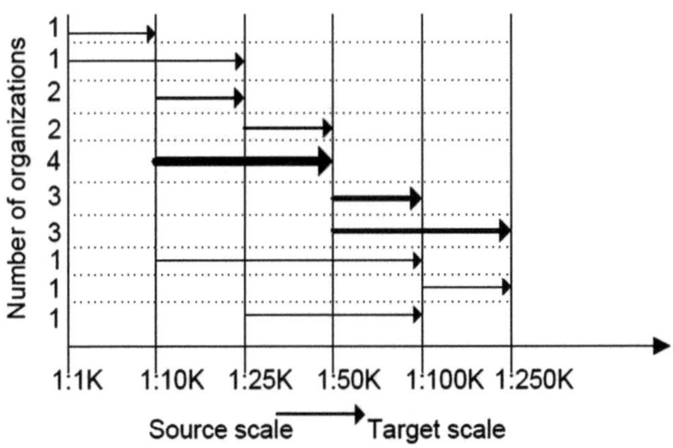

Fig. 3 Overview of the different scale transitions applied at NMAs [adopted from Foerster and Stoter (2008)]

Germany mentioned in (ATKIS 2005) in the project Authoritative Topographic-Cartographic Information System (ATKIS) (Stoter 2005). The generalization model used in the German ATKIS project represents a different view of the overall generalization process with an addition of *object generalization* along with object and cartographic generalization (Steiniger and Weibel 2005). Initially, data from real world is transformed to Digital Landscape Model (DLM) with scale 1:25,000 based on newly developed object generalization. Subsequently, DLM50 and Digital Cartographic Model (DCM) 25 are derived based on model and cartographic generalization concepts respectively. CHANGE for buildings generalization, ESRI/AED-SICAD/Open Software is applied for generalization purposes.

In order to understand the implementation of new technology to automate map generalization and incremental updating processes in France (IGN), we focused on the research conducted by Jahard et al. (2003) to derive 1:100,000 scales of series of map from one database called *BDCarto® database* for whole French territory. According to their literature, introduction of new technology helped reduce cost, operator's work, and duration. This is evident from the two projects launched in 1994 and with introduction of new technology in 1999. The generalization process had been estimated to 1,000 h of operator work per map in 1994 compared to about 50 h on a Pentium III computer with addition of about 100 h as interactive part in 1999. While the whole process took 16 months (2,500 h) in 1999 thus, with new technology, the productivity is increased 10 times more than the previous project.

Generalization process is not applied to all databases at Spain (ICC). However, out of the remaining databases, some maps are generalized automatically, while others are derived with the help of manual alternation or editing. Map 10 M is generalized from BT-5 M database using manual editing while Map 5 and Map 25 M are generalized automatically. Similarly, generalization of map from another map is also part of generalization process in ICC (Spain) (ICC 2012). For example, Map 100 M is generated from Map a *Comarcal* map. The process of the generalization of maps is being improved with the help of self developed algorithms in "Clarity" at Denmark—Kort and Matrikelstyrelsen KMS (Denmark) (KMS 2012). The output obtained from automation process is verified manually and modifications made. The dataset with 1:100 k scales is maintained separately from the other existing datasets because of six years updating cycle.

In Swiss NMA (Swisstopo 2012), "*In the current process the generalization, which is performed to propagate updates, requires 100 % human interaction. Swisstopo is at the beginning of a major change in which the whole data management including generalization will be reconsidered*" (Stoter 2005). Derivation multiple-scale maps and tools used to implement generalization processes at a number of NMAs are given in Table 1.

A detailed discussion related to generalization tests performed within European *Organization for Experimental Photogrammetric Research* (OEEPE; the predecessor of Euro-SDR) to evaluate the results of different interactive generalization software is made. These tests were specific to test datasets, generalization operations, cartographic conflicts, and algorithms. The potential capacities of several

Table 1 Tools used to derive multiple-scale maps at NMAs

NMA	Generalization (from–to)	Tools used
IGN, France	BDTopo–1:25 k map	DataDraw (self-developed)
	BDCarto–1:100 and 120 k	Gelbe (self for roads)
	BDTopo–1:500 k map	LAMPS2/Agent and ArcInfo (for roads)
OS, Great Britain	MasterMap–1:25, 1:50, 1:250 k	ESRI Software, Laser-Scane (Intergraph), MapInfo and Clarity (Laser-Scan)
TDK, The Netherlands	1:100–1:250 and 1:250 k	MicroStation
Catalonia, Spain	BT-5 M–1:5 k map	CHANGE for buildings
	BT-5 M–1:10 k map	CHANGE(automatic)
	BT-25 M–1:25 k m	CHANGE + ICC + interactive
IGN, Belgium	1:100 k (Top100v)–1:50 k	ArcGIS, AML, Laser-Scan
	1:10–1:50 and 1:100 k (in-process, (Stoter 2005)	Lamps2/Agent
BKG, Germany	DGK5–DLMB	CHANGE for buildings generalization, ESRI/AED-SICAD/Open Software and Oracle
	JOG250–DLM250	
	IWK1000–DLM1000	
	DLM B–DLM50, DSK10, DTK25	
	DLM 50–DTK50, DTK100	
	DLM250–DTK250	
	DTK25–TK25	
	DTK50–TK50	
	DTK100–TK100	
	DTK250–TUK250	
	DTK500–UK500	
	DTK1000–IWK1000	
	100, 250, 500, 1,000)–	
	TK 25, 50, 100, 250, 500, 1,000)	
GCM, Turkey	1:50–1:100 k	LAMPS2 By "Laser-Scan" and Lull (Programming language)

generalization programs on selection of test cases are tested within EuroSDR research project. Additionally, a good literature about the main findings and conclusions of state-of-the-art of automated generalization in commercial software can be found (EuroSDR 2010).

3 A Conceptual Framework for Generalization of Maps

A number of frameworks for generalization of map have been proposed over the past three decades. These frameworks can be categorized based on two correlated models: process-oriented generalization models and object-level-oriented generalization models related to level of map objects (Steiniger and Weibel 2005). Frameworks proposed by Brassel and Weibel (1988) and McMaster and Shea (1992) can be considered as process-oriented generalization models while other frameworks proposed in (Ruas and Plazanet 1996) and its enhanced version in the AGENT project (Ruas 1999) as object-level-oriented generalization models. Similarly, a framework proposed in (Steiniger and Weibel 2005) is based on three processing stages: structure analysis, generalization and visualization with respect to thematic mapping.

The proposed framework presented in Fig. 4 is based on a combination of both models mentioned above for map and database generalization similar to Brassel and Weibel (1988), who distinguished generalization processes into five steps: (1) Structure recognition, (2) process recognition, (3) process modeling, (4) process execution, and (5) data display. This framework is composed of five (5) steps: (1) Recognition of structure of database or map composed of identification of objects or aggregates, spatial relations and priority or importance, control by target-scale and generalization rules as generalization objectives. The output of this process is passed-on to the next step. The types of data modification and the parameters of the target structures (map or database) have to be established to determine what is to be done with the original map or database based on the control parameters. (2) Essential features from structured spatial data are recognized based on the parameters mentioned above, which include different types of conflicts, structure of target-database etc. (3) Compilation of rules and procedures from an operation library and the prior-setting of parameters of operations, made part of operation modeling. Original database or map and structure of information depend on this operation to be converted into the target generalized map or database. (4) Operation *execution* consists of a sequence of operational steps compiled from generalization functions stored in the operations library. Generalization functions can be selection or line simplification, elimination, displacement and feature aggregation. (5) *Data display step transforms* the target data into the target generalized map or database.

The goal of automated generalization within this framework is to find a good compromise between all these several constraints. Thus, a minimal acceptability or compromising factor between all constraints must be found during data

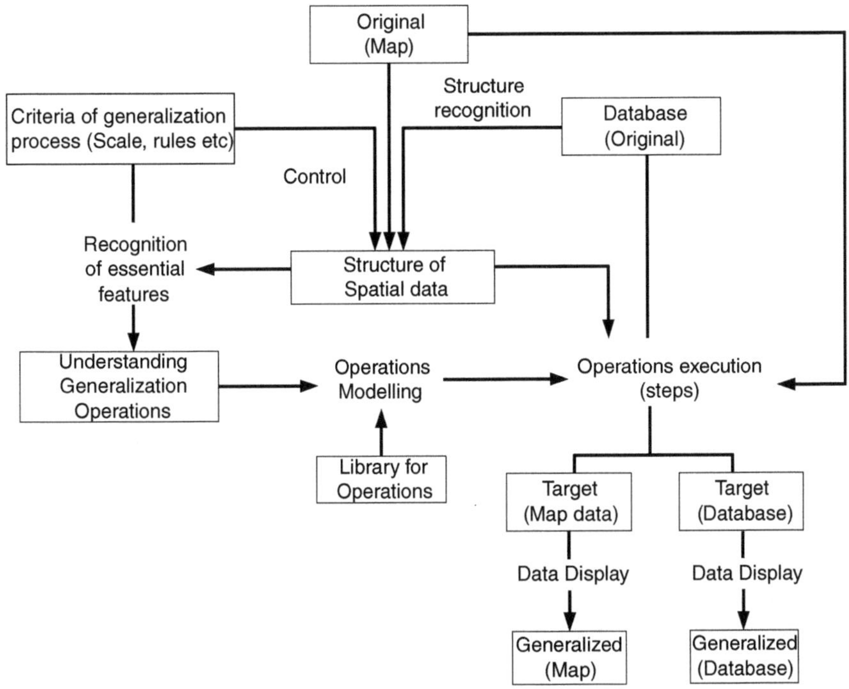

Fig. 4 Conceptual framework for generalization of maps and geo-database

conversion. This can be expressed by a compromising function that is, as for our experiments, just the simple average of all constraint values (Neun et al. 2009). In order to favor and punish certain constraints a weighting can be applied to them when calculating the compromising factor function.

Compromising Factor (CF) = $\sum$ (Constraint (a, b, c, d….h) *Weight)/n (Neun et al. 2009).

- Where alphabets (a, b, c) = constraints
- Value of "weight" is between scales [0–1]. The lowest value 0 means, no violation of constraints and 1 means maximum violation of constraints. "n" represents total number of constraints.

The constraints and their weights are the only parameters that can be adjusted in order to modify the current system. Operator sequences are created by iteratively executing all the available generalization operators and then selecting one or more results using a heuristic in order to continue the process, again executing all operators. The search algorithms used in our research are hill climbing, simulated annealing and genetic search (Neun et al. 2009). With a set of operators there exists a large space of possible operator sequences which can be pursued iteratively.

4 Implementation and Results

The platform used for the generalization process was ArcGIS software from ESRI products operated on a Core (TM) i7 CPU 2.40 GHz, 4.00 GB RAM. A portion of this framework is implemented and tested on datasets of *Kelantan* state in Malaysia with the aim to producing generalized maps.

Generalization operators influence the outcome of the generalization processes heavily and are one of the building blocks for automated generalization (Foerster and Stoter 2008). A number of generalization operations were conducted as part of this research: simplification of polygons, and transportation network (roads), aggregation of simplified polygons and smoothing of polygons. The data used for this purpose was a small area of *Kelantan* state in Malaysia with dimensions 6.5 km by 3.5 km as shown in Fig. 5. Input for simplification process and aggregation operations was a set of 7,014 polygons in *Kelantan*. Input data for line simplification is a set of transportation network (roads) of *Kelantan* presented in Fig. 6.

Polygons are simplified based on two simplification algorithms: *removal of points* and *bend simplification* by applying threshold values for maximum allowable offset and minimum area to be considered for simplification. The result of this process is presented in Fig. 7a, b. A map of polygons by applying first algorithm based on *removal of points* is presented in Fig. 7a (red) while bend simplification in Fig. 7b (blue). Complete structures of simplified polygons could not be maintained due to removal of vertexes (points) randomly compared to the

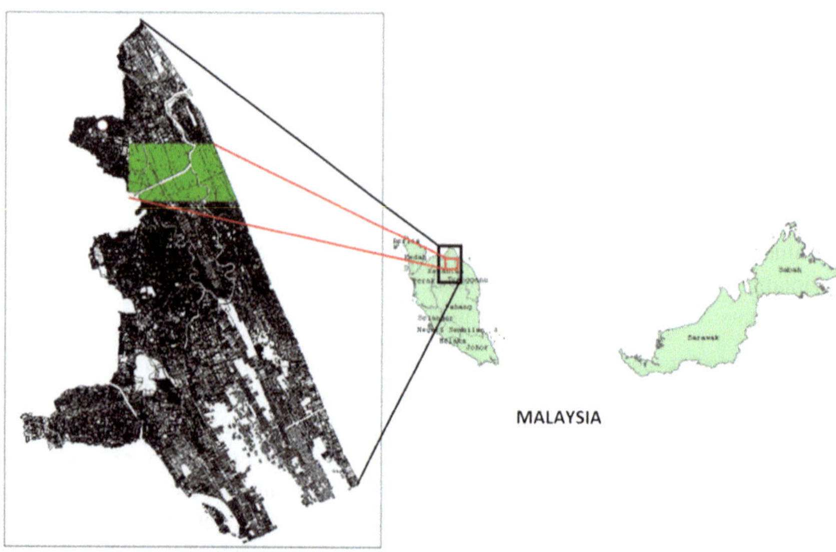

Fig. 5 An area of 6.5 by 3.5 km in Kelantan state, with 7,014 polygons (*green*)

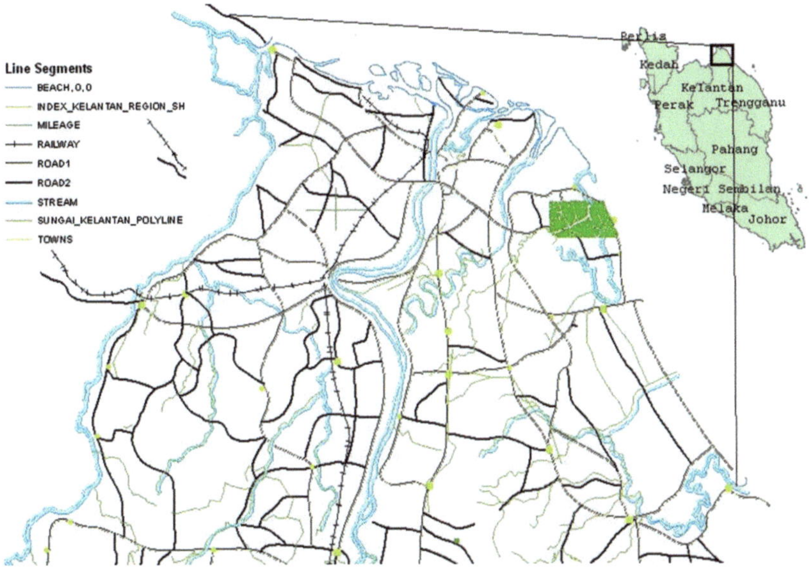

Fig. 6 A set of transportation network (roads and railway lines), rivers, streams in Kelantan state, Malaysia

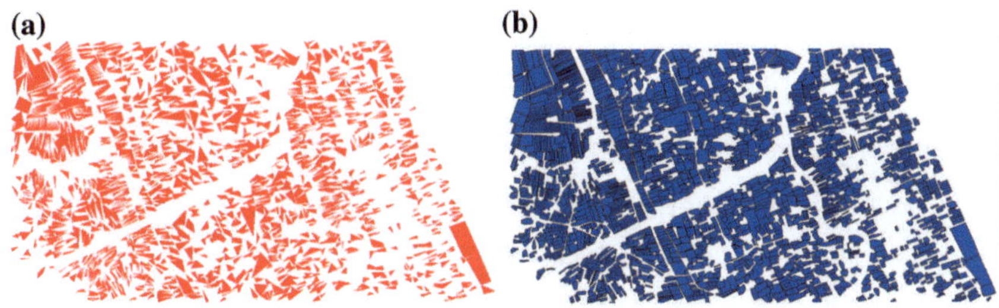

Fig. 7 **a** Simplified polygons based on point removal algorithm and **b** bend algorithm (*right*)

bend simplification. The processing time for simplification was higher in *bend simplification* as compared to *removal of point simplification.*

Simplified polygons are aggregated based on the following parameters: distance between parcels, minimum area, minimum hold area and preservation of orthogonal area. The threshold values for each parameter were provided to estimate time collapse and analyzed. The result of the aggregation process is presented in Fig. 8. A map of aggregated polygons based on distance between parcels (10, 50 m), minimum area (5, 10 m) minimum hold area (0, 2 m) with and without preserving orthogonal area in Fig. 8a, b. The processing time for aggregation for all parameter values was more or less same with slight difference in seconds.

Line simplification process was applied to simplify line features such as roads unlike polygon simplification. Two algorithms were applied for this purpose:

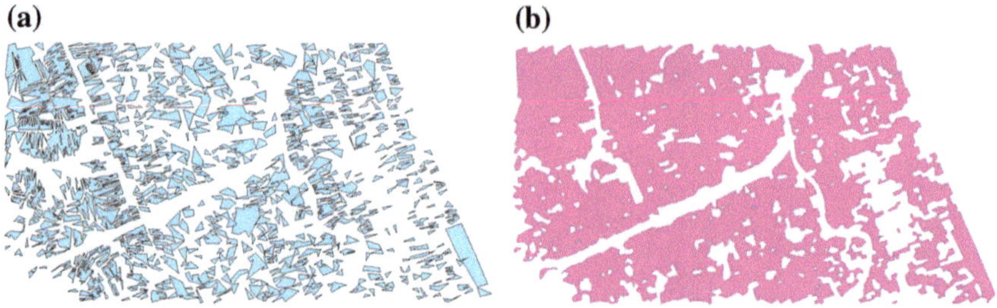

Fig. 8 Aggregation of simplified polygons based on two different values of parameters (100 m) shown in (**a**), and (500 m) in (**b**)

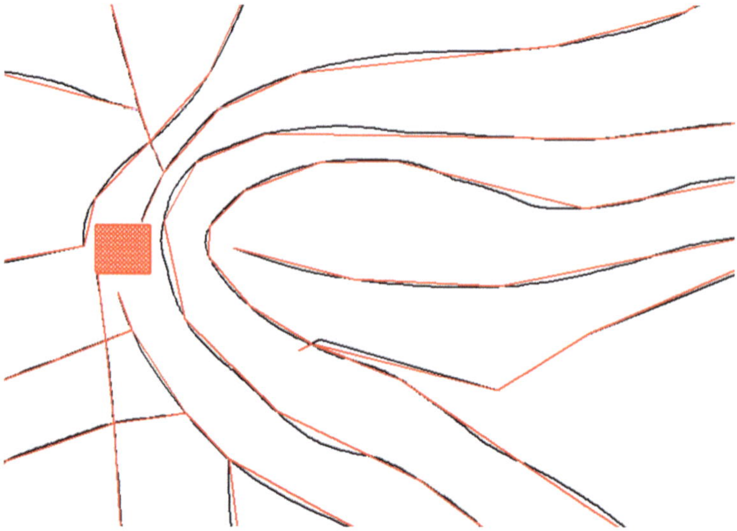

Fig. 9 Simplified transportation network (roads): original roads (*black*) and simplified roads (*red*)

simplification based on removal of points and bend simplification. Different parameter values for simplification tolerance were applied. The result of line simplification method is presented in Figs. 9 and 10. Line features (black) are original while simplified lines are illustrated with (red) color in Fig. 9.

Two smoothing algorithms (Paek and Bezier interpolation) were applied providing values for tolerance and preserving end points for rings. The result of smoothing of polygons is presented in Fig. 11.

JUPEM maintains three types of data products: cadastral, mapping and geodetic products. Cadastral products include digital lot, certified plan; standard sheets while mapping products include restricted and unrestricted topographic maps (JUPEM 2012). Seven series of unrestricted topographic maps are available

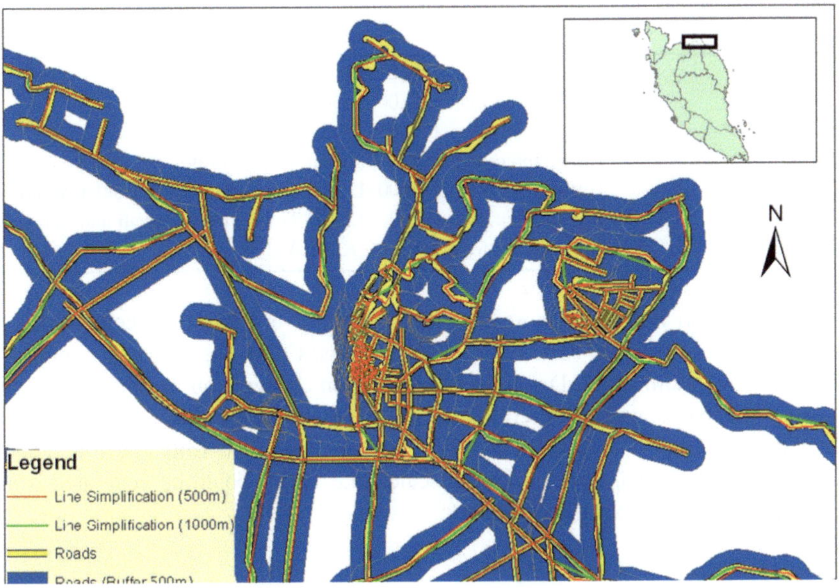

Fig. 10 A buffer zone is created around roads along with simplified transportation network (roads): original roads (*yellow*), simplified with different tolerance (*red* and *green*)

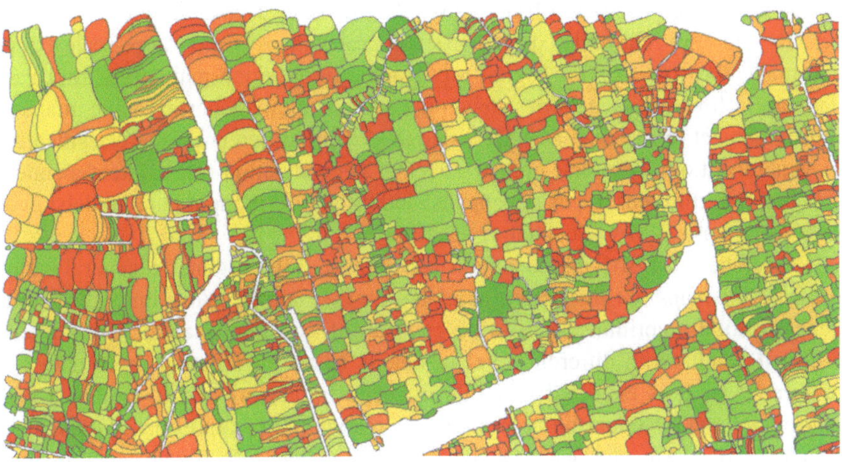

Fig. 11 Smoothed polygons of a small portion of Kelantan state, Malaysia

presented in Table 2. Data about benchmarks, GPS stations, gravity, tides and *Geoidal* maps are also currently available with *JUPEM*. All these products are available at different scales. Therefore, implementation of the above-mentioned framework in NMAs especially *JUPEM*, Malaysia for generalization of topographic data could be beneficial.

Table 2 Unrestricted topographic maps (JUPEM 2012)

Series no.	Topographic maps
DNMM5101	Topographic map 1:50,000 Peninsular Malaysia
DNMM5210	Topographic map 1:50,000, Sabah and Sarawak
DNMM6102	Topographic map 1:25,000 Peninsular Malaysia
DNMM6201	Topographic map 1:25,000 Sabah and Sarawak
DNMM8101	Topographic map 1:3,000–12,500 Town Map in Peninsular Malaysia
DNMM8201	Topographic map 1:3,000–12,500 Town Map in Sabah and Sarawak
DNMM6101	Topographic Map 1:10,000 Klang Valley

JUPEM also maintains more than four series of topographic vector datasets such as T 735, T 738, T 834, T 835. Four databases from these datasets could be produced and maintained. Database generalization concept can be applied and six smaller databases with different scale like GeoDb 10, GeoDb 25, GeoDb 50, GeoDb 100, GeoDb 250, GeoDb 1000 could be generated from the four major databases. Digital maps with different scales like 1:10 k could be generated from GeoDb 10, 1:25 k from GeoDb 25 etc.

5 Summary

In regard with generalization of cartographic maps, individually, some of European NMAs such as BKG of Germany, IGN of France, and TDK of The Netherlands etc. are currently deriving small-scale maps from large-scale maps automatically from existing digital topographic information maintained in a geodatabase. However, a generic view on generalization of cartographic maps is mostly subject to research at NMAs due to existing differences: the classification of generalization operators; mutual effect of generalization rules; interdependencies between different (types of) objects; and difference in input data models; even within the same NMA (Foerster et al. 2010).

Researchers are striving to obtain a complete fully automatic generalization package for maps but this is not available on the market. Work on task-specific generalization algorithms need to be carried out as different applications intend different results with different level of details based on different input data and requirements. It is expected that the proposed framework would enable a user (cartographers) the opportunity to define the relevance of features based on distinct characteristics. An interface can help channel all the specialized generalization procedures for successful completion of the intended process. A customized interface module for generalization algorithms will explicitly state what kind of datasets and features they operate and for which map scale. Relevant tools for interface, parameterized generalization algorithms are proposed for application dependent generalization operators for all features. This includes a generalization module for conflict resolution of different generalization approaches, programming methodology and model for simplification of features.

Currently, different applications demand different abstractions or levels-of-details (LoDs) of 3D city models because city models contain rich geometric and semantic information which effects efficient processing and vitalization. Therefore, as part of future work, lower LoDs from higher LoDs of 3D city models would be generated by reducing data volume.

Acknowledgments We would like to convey our deepest acknowledgement firstly to Universiti Teknologi Malaysia for providing Research Grant Project Vote No. R.J130000.2427.00G24, *An Innovative 3D Geo DBMS for Asset and Facilities Management.* Additioanlly, our sincere appreciations to Research Management Centre (RMC) of UTM and Ministry of Higher Education (MOHE), Malaysia for enabling us to carry out this research project. We would like to thank Nur Amalina Zulkifli for providing the Kelantan state dataset.

References

ATKIS (2005) Amtliches Topographisch-Kartographisches Informationssystem. www.atkis.de. Accessed 13 Aug 2011

Brassel KE, Weibel R (1988) A review and conceptual framework of automated map generalization. Int J Geogr Inf Syst 2(3):229–244

EuroSDR (2010) EuroSDR research on state-of-the-art of automated generalisation in commercial software: main findings and conclusions. 13th Workshop of the ICA commission on Generalisation and Multiple Representation, Zurich

Foerster T, Stoter J, Kraak MJ (2010) Challenges for automated generalisation at European mapping agencies: a qualitative and quantitative analysis. Cartographic J 47(1):41–54

Foerster T, Stoter JE (2008) Generalisation operators for practice: a survey at national mapping agencies. In: 11th ICA workshop on generalisation and multiple representation, Montpellier, 20–21 June 2008, p 11

Grünreich D, Powitz B, Schmidt C (1992) Research and development in computer-assisted generalization of topographic information. In: Third European conference on geographical information systems, Institute of Cartography, Hanover University, Hanover, pp 532–541

ICC (2012) Institut Cartogràfic de Catalunya. www.icc.es. Accessed 30 April 2012

Jahard Y, Lemarié C, Lecordix F (2003) The implementation of new technology to automate map generalisation and incremental updating processes. In: 21st International Cartographic Conference (ICC), Durban, South Africa, 10–16 Aug 2003. Document Transformation Technologies, pp 1449–1459

JUPEM (2012) JUPEM Geoportal. http://www.jupem.gov.my/index.php. Accessed 13 Jan 2012

Kazemi S, Lim S, Linlin G (2007) An international reseach survey: cartographic generalisation practices at mapping agencies. In: Spatial Science Institute Biennial International Conference, Hobart, Tasmania, Australia, 14–18 May 2007, pp 537–556

Kilpeläinen T (2000) Knowledge acquisition for generalization rules. Cartography Geogr Inf Sci 27(1):41–50

KMS (2012) Kort and Matrikelstyrelsen (Denmark). www.kms.dk. Accessed 15 July 2012

Lecordix F, Gallic JL, Gondol L, Braun A (2007) Development of a new generalization flowline for topographic maps. Paper presented at the 10th ICA workshop on Generalisation and Multiple Representation, Moscow, Russia

Li Z (2007) Digital map generalization at the age of enlightenment: a review of the first forty years. Cartographic J 44(1):80–93

Mackaness WA, Ruas A, Sarjakoski LT (2007) Generalisation of geographic information: cartographic modelling and applications. International Cartographic Association, 1st edn. Elsevier, Amsterdam
McMaster BR, Shea SK (1992) Generalization in digital cartography. Association of American Geographers, Washington
Neun M, Burghardt D, Weibel R (2009) Automated processing for map generalization using web services. GeoInformatica 13(4):425–452
Perkal JD (1966) An attempt at objective generalisation (1958) (trans: Jackowshi W). Department of Geography, University of Michigan, Michigan, USA
Regnauld N, Revell P (2007) Automatic amalgamation of buildings for producing Ordnance Survey® 1:50,000 scale maps. Cartographic J 44(3):239–250
Ruas A (1999) Modèle de généralisation de données géographiques à base de contraintes et d'autonomie. Université de Marne-la-Vallée
Ruas A, Plazanet C (1996) Strategies for automated generalization. In: Kraak MJ, Molenaar M (eds) Advances in GIS Research II (7th International Symposium on Spatial Data Handling), London, 1996. Taylor and Francis, pp 6.1–6.18
Sester M (2002) Application dependent generalization: the case of pedestrian navigation. Int Arch Photogrammetry, Remote Sens Spat Inf Sci 34(4):291–296
Steiniger S, Weibel R (2005) A conceptual framework for automated genealization and its application to geologic and soil maps. Paper presented at the The International Cartographic Association (ICA-ACI), A Coruña, Spain, 11–16 July
Stoter JE (2005) Generalisation within NMA's in the 21st century. Paper presented at the International Cartographic Conference
Swisstopo (2012) swisstopo, the Federal Geo-Information center. http://www.swisstopo.admin.ch/internet/swisstopo/en/home.html. Jan 16
Tobler WR (1966) Numerical map generalization; and, Notes on the analysis of geographical distributions. Department of Geography, University of Michigan, Ann Arbor, Mich
Töpfer F, Pillewizer W (1966) The principles of selection. Cartographic J 3(1):10–16
Weibel R, Dutton G (1999) Generalizing spatial data and dealing with multiple representations. In: Longley P, Goodchild MF, Maguire DJ, Rhind DW (eds) Geographical information systems: principles, techniques, Management and Applications, Chichester, Wiley, NY, pp 125–155

Methods for Georeferencing Point Cloud of Building From Static TLS: A Review

Rizka Akmalia, Halim Setan, Zulkepli Majid and Deni Suwardhi

Abstract Today, 3D model of building has been used by many applications. In the other hand, the popularity of Terrestrial Laser Scanner (TLS) has also increased significantly since it can capture 3D object directly by scanning the object. However, its application and data processing have been into specific issues such as georeferencing the point cloud. Georeferencing is a process of transforming the TLS data coordinate system into a world coordinate system. Several TLS applications require this process, especially a project that consist of large objects such as buildings; in order to analyze it as a part of a certain environment, or as a step to combine TLS data with other measurement data into universal coordinate system. In this chapter, an overview and comparison from some previous works about TLS data georeferencing, especially for buildings, are discussed. Some improvements that have been made from previous works in georeferencing techniques on the field as well as processing in the laboratory are also included.

Keywords Georeferencing · TLS · Building application · Point cloud transformation · 3D model

1 Introduction

Three dimensional (3D) modeling has become more popular and the demand of 3D data has also increasing (Aydar et al. 2011; Yildiz and Gumusay 2011; Dursun et al. 2008; Voegtle and Steinle 2000). 3D City Model, which has been developed

R. Akmalia (✉) · H. Setan · Z. Majid
Faculty of Geoinformation and Real Estate, Universiti Teknologi Malaysia, Johor Bahru, Malaysia
e-mail: rizka.akmalia01@gmail.com

D. Suwardhi
Faculty of Earth Science and Technology, Bandung Institute of Technology, Bandung, Indonesia

A. Abdul Rahman et al. (eds.), *Developments in Multidimensional Spatial Data Models*, Lecture Notes in Geoinformation and Cartography,
DOI: 10.1007/978-3-642-36379-5_13,

within this past years, has being used by many sectors. Building, as one of the vital object in urban development, becomes an important object to be modeled in 3D, especially in 3D City model (Sester 2007; Yang et al. 2002; Doellner and Bucholz 2005). Several applications such as architectural, urban planning, administration, facility management, and engineering construction use 3D model of building as an essential data for further analysis. 3D model of building can give important information and even give a contribution for energy demand forecasting (Strzalka et al. 2011). Thus, the availability of a detail and accurate 3D model of building is needed.

Nowadays, instead of using photogrammetry, active sensors are usually used to generate 3D models of objects and sites (Guidi et al. 2008). A traditional method like photogrammetric image analysis was time consuming task (Voegtle and Steinle 2000; Lee et al. 2009). Using aerial active sensor like Light Detection and Ranging (LIDAR) can capture a large area but many of object details will be missed (Bohm 2005). Due to its practicality and versatility, TLS has been used for most building applications instead of using any other active sensor (Bornaz and Rinaudo 2005; Susaki 2009; Altuntas et al. 2010; Boulaassal et al. 2011). TLS can be used to capture the details with high precision in a relative short time scanning. Also, TLS enables digital documentation of buildings, sites and physical objects for reconstruction and restoration including cultural heritage (Vela et al. 2010; Baz et al. 2008; Shi et al. 2009). Many advantages in a process to generate 3D model of building can be achieved by using TLS.

Objects scanned by TLS were usually referred to the local coordinate system. A process to transform the scan coordinate system into geodetic coordinate system is called georeferencing. For building application, which is usually captured using several sensors e.g., photogrammetry, laser scanning, and LIDAR, georeferencing is important as a frame to combine all the data in a certain coordinate system. It integrates TLS data and the products derived from it, e.g., 3D models, with other geospatial data. Not all applications need georeferencing process, but most of architectural, archeological conservation, engineering, urban planner, and constructional applications need this process to see the model position in a world coordinate system. Thus, it needs to georeference the model into a certain world coordinate system.

The requirement of georeferencing accuracy for building modeling which been used for architectural purposes is usually enough for 5–10 cm (Scaioni 2005). In other fields like heritage documentation, a complete modeling of a large site is more important than the highest accuracy in georeferencing (Alba et al. 2007). Usually, to do georeference, other measurement tools will be integrated with TLS such as total station or GPS.

The highlighted problem in the georeferencing is how to do it with a minimum effort but with high accuracy. This chapter reviews current methods in georeferencing and also the implementation in study cases from previous works will be described. Thus, the most suitable method to apply on their cases can be considered.

2 Coordinate System in TLS

TLS measures 3D object by scanning the object; obtaining thousands of points represent the object's surface which is called point cloud. A laser scanner usually can perform to scan a large object resulting point cloud up to 12k points per second (Alba et al. 2007). Points coordinates can be referred to the local coordinate system, or certain coordinate system. Generally, every point in point clouds was referenced to the internal scanner coordinate system or instrument-fixed. Due to the differences in laser scanner design, every laser scanner manufacturer has its own definition for instrumental-fixed coordinate system.

Scanning a large object such as building, will take some scan positions to capture all of object's surfaces. Since each scanning position has its own scanner coordinate systems, to see the model as a complete object, it is required to combine all of the scanned parts. Actually, to combine all scanned parts, registering process is enough. But, using this process, the coordinate system will still remain in scanner coordinate system. In building application, user usually needs to see the model on its position in the earth to analyse the model within the environment. Thus, transformation process from scanner coordinate system into real world coordinate system or external coordinate system is needed.

Transformation from scanner coordinate system to the external coordinate system needs six parameters to be solved, i.e. ΔX, ΔY, ΔZ, ω, φ, κ, and often called Exterior Orientation Parameters (EOPs). To solve it, minimum 3 control points, known for both in scanner coordinate and external coordinate, are needed. Figure 1 shows the coordinate system between scanner system and external coordinate system.

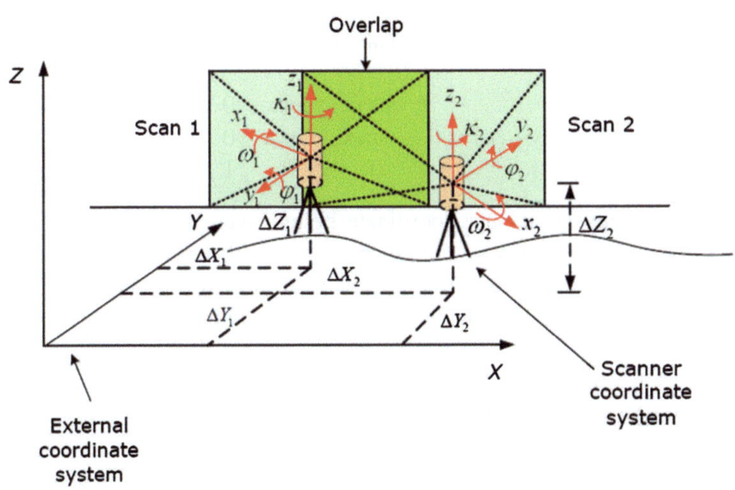

Fig. 1 Difference between scanner coordinate system and external coordinate system [*source* (Reshetyuk 2009)]

This transformation process is called georeferencing. There were several methods to do georeferencing in practice. However, theoretically, the process is to solve the translation parameters (ΔX, ΔY, ΔZ) and rotation parameters (ω, φ, κ) (Alba et al. 2007; Karabork et al. 2012; Scaioni 2005). There is no scaling factor because TLS works similarly to Electronic Distance Measurement (EDM). It measures the distance for every point in the exact scale in the real world. Since there is no scaling in georeferencing from point cloud, it is often called as rigid-body transformation. While the scanner coordinate system usually centered in the instrument fixed and follow the right-hand rules for the direction of xyz, external coordinate system follow the chosen reference for the direction of xyz (e.g., WGS 84, GDM 2000, etc.).

If the scanner coordinate system has a notation of $\boldsymbol{X_s}$ and external coordinate system has a notation of $\boldsymbol{X_e}$, then the translation vector from scanner coordinate system to external coordinate system is $\boldsymbol{\Delta X_{se}}$. Also, the rotation vector from the scanner coordinate system into external coordinate system is $\boldsymbol{R_{se}}$. Mathematically, Eqs. 1, 2, and 3 can be used to calculate the coordinate in external coordinate system.

$$\boldsymbol{X}_e = \boldsymbol{\Delta X}_{se} + \boldsymbol{R}_{se}\,\boldsymbol{X}_s \tag{1}$$

$$\Delta X_{se} = [\Delta X_{se}\,\Delta Y_{se}\,\Delta Z_{se}]^T \tag{2}$$

$$\boldsymbol{R}_{se} = R_3(\kappa).R_2(\varphi).R_1(\omega) \tag{3}$$

$$R_1(\omega) = \begin{pmatrix} 1 & 0 & 0 \\ 0 & \cos\omega & \sin\omega \\ 0 & -\sin\omega & \cos\omega \end{pmatrix}$$

$$R_2(\varphi) = \begin{pmatrix} \cos\varphi & 0 & -\sin\varphi \\ 0 & 1 & 0 \\ \sin\varphi & 0 & \cos\varphi \end{pmatrix}$$

$$R_{(3)}(\kappa) = \begin{pmatrix} \cos\kappa & \sin\kappa & 0 \\ -\sin\kappa & \cos\kappa & 0 \\ 0 & 0 & 1 \end{pmatrix}$$

Points coordinate in scanner coordinate system were known, while points in exterior coordinate system were unknown. To solve the unknown, control points were needed. Additional measurement tools were needed to measure the control point position in external coordinate system or for the orientation of scanner.

3 Classification of Georeferencing Methods

There are two methods to conduct georeferencing using static TLS, i.e., direct method and indirect method. The different between those two methods is in the processing to transform the coordinate whether the transformation is calculated in the field directly or after the data has been transferred to the computer. For mobile TLS, usually only direct method is used for georeferencing.

Traditionally, georeferencing process will be conducted indirectly, using control points which position is already known in local and world coordinate system. To measure the position of control point in the field, another measurement tools like total station will be used in the field. Afterward, points will be used to solve the EOP and then all of point cloud will be transformed into world coordinate system using the solved parameter. However, technically, there are some methods in the field to do georeferencing process. Figure 2 shows methods which are usually used to do georeferencing.

Georeferencing using mobile TLS were conducted by integrating Global Positioning System (GPS), Inertial Measurement Unit (IMU), and TLS. In the process, GPS receiver will receive data from satellite and directly measure the position of laser in World Geodetic System (WGS) '84 coordinate system. The laser orientation will be measured using IMU. The georeferencing process is conducted directly in the field. Thus, this method is called as a direct method. Figure 3 shows the laser scanner mounted on a vehicle for mobile scanning. Usually the GPS and IMU will be attached on the scanner.

For most building application, static TLS will be used. There are several techniques to do georeferencing using static TLS, i.e., control point, sensor driven, and data driven. Those techniques can be grouped as direct and indirect methods. Each technique has its own advantages and advantages. Thus, it should be considered which technique is the most suitable according to the requirement.

3.1 Direct Method

Direct georeferencing method means that the TLS scans the object from a known point coordinate. This method is suitable for cases where the positioning of

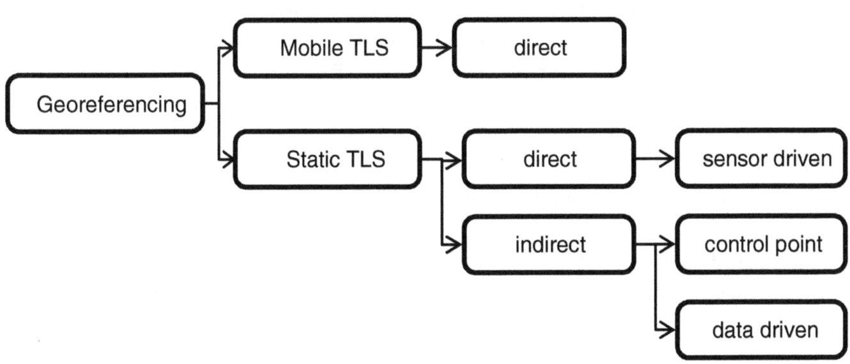

Fig. 2 Georeferencing methods

Fig. 3 Mobile TLS application [*source* (Harrap and Lato 2010)]

Ground Control Point (GCP) is not possible. To place the TLS over a known point coordinate, the use of GPS or the existing benchmark is needed. Digital compass or IMU can be used for to solve the orientation of TLS. Then, the position and orientation information where the TLS has stand can be inserted into the scanner system. Thus, all resulted point clouds are transformed directly into the system coordinate used as a reference to measure the scanner coordinate.

Advantages of using this method are save the money and time since no GCP or overlap scans are needed in this method. However, this method gives lower accuracy than indirect method. For application requiring medium accuracy and short time in data acquisition and processing, this is the most suitable method.

Using direct method with sensor driven approach, the method and additional tools for positioning survey can be chosen. Sensor driven technique for static TLS actually was adapted from the aerial scanning. The concept is the same that GPS will be used to measure the sensor position. If the aerial scanning or mobile TLS needs high-end GPS plus IMU to meet the accuracy, static TLS only requires low-cost receiver to achieve the adequate accuracy. Using this technique, the receiver will be easily attached on the TLS, shown in Fig. 4, and no need to place any target around the object. Automation in data processing in this technique is also possible.

The use of GPS also varied from static GPS into RTK. The accuracy of georeferencing using this approach will be determined by the method of GPS signal recipients. The accuracy resulted from different method will be different. Static GPS can give a better accuracy than RTK. However, some experiments show that some improvement in this method can give many impacts in the accuracy.

The use of low-cost GPS has been tested in the Schillerplatz experiment. This experiment was done by (Schuhmacher and Bohm 2005) using Leica HDS 3000 with Real-time low-cost GPS. The purpose of this experiment was for modeling the façade of some historical buildings in that area. Result of accuracy was about 4 m for 20 min signal recorded for every scan station.

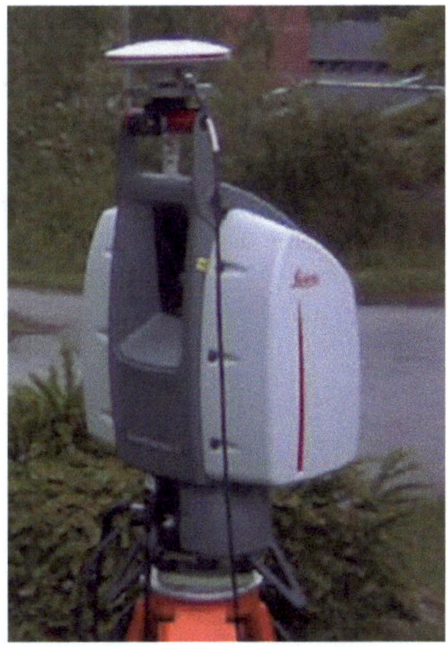

Fig. 4 TLS with attached antenna GPS on top [*source* (Reshetyuk 2009)]

Experiment using low-cost GPS was also been tested by Schuhmacher and Bohm (2005) for Schloss Rosenstein palace. They used traverse method with 5 points as the point to stand TLS. Low-cost GPS and digital compass were used to measure the position and orientation of each station. Result from this experiment was not so much different with the Schillerplatz experiment. The horizontal accuracy still larger than the vertical because the additional error from digital compass.

The improvement of accuracy was achieved by the experiment from (Karabork et al. 2012) and (Reshetyuk 2009). Karabork et al. 2012 achieved the accuracy in direct georeferencing using GPS less than 10 cm. Great improvement was made by Reshetyuk (2009) achieving accuracy into millimeter. Static GPS with two GPS antennas, one mounted on the scanner, and another in another point, were used in the experiment. RTK GPS with variation in receiving time from 5 into 10 min were also tried. The result had shown that the longer the antenna receive the GPS signal, the better the accuracy. Table 1 shows the result from previous works using direct georeferencing methods.

3.2 Indirect Method

Indirect method means that the point cloud is transformed after the scanning process. There are two approaches in indirect methods i.e., control point and data driven. Using control point is the most suitable technique for applications requiring high accuracy in georeferencing. However, for the automation in data process preferably, data driven is still the most suitable technique.

Table 1 Results from previous works on direct georeferencing

Project	Georeferencing technique	Accuracy (m)
Shillerplatz by Schuhmacher and Bohm (2005)	RTK low-cost GPS	X 2.8 Y 2.7 Z 0.1
Schloss Rosenstein by Schuhmacher and Bohm (2005)	RTK low-cost GPS and digital compass	X 2.3 Y 2.7 Z 2.0
Resetyuk (2009)	Static GPS	X 0.048 Y 0.056 Z 0.056
	RTK GPS (10 min)	X 0.075 Y 0.075 Z 0.15
	RTK GPS (5 min)	X 0.12 Y 0.083 Z 0.045
	Traverse (scanning from known coordinate)	X 0.005 Y 0.007 Z 0.0035
Scaioni (2005)	Traverse (with distance constraint from TLS to object not larger than 55 m)	X 0.039 Y 0.031 Z 0.025
Alba et al. (2007)	Traverse	XY: 0.09 Z: 0.035
Paffenholz and Kutterer (2008)	RTK GPS	Range: 0.05–0.1
Karabork et al. (2012)	RTK GPS (5 scan stations)	X 0.78 Y 0.72 Z 0.75
	RTK GPS (6 scan stations)	X 0.18 Y 0.167 Z 0.176

Control point technique applies total station to measure the position of control points in the field. This technique is also called as indirect method since georeferencing process will be done off-field. Practically, several special targets (minimum 3 targets) usually will be placed around scanned object as control points. It is the very common technique in georeferencing.

Advantages of using control point technique are the high accuracy, no need to center the TLS upon certain point, and no need to measure the instrument height. Though, using this method will be time and cost consuming since it needs secondary measurement tools e.g., Total Station and target to be placed as control point. Several experiments had proved that resulted accuracy from this technique is less than 1 cm (Vozikis et al. 2004). Thus, for application requiring high such as

Table 2 Results from previous works on georeferencing using control point

Project	Georeferencing technique	Accuracy (m)
Schuhmacher and Bohm (2005)	Control point using total station as additional measurement tools	X 0.001 Y 0.002 Y 0.002
Alba et al. (2007)	Control point (with total station)	XY: 0.05 Z: 0.03
Vozikis et al. (2004)	Control point (with total station)	Range = 0.01–0.02
Gonzales-Aguilera et al. (2008)	Control point (with total station)	Average = 0.003
Alba et al. (2006)	Control point (with total station)	XY: 0.002 Z: 0.003
Miri and Varshosaz (2011)	Control point	Range = 0.025–0.05

deformation detection (Alba et al. 2006; Gonzales-Aguilera et al. 2008), control point technique is the most suitable approach.

Some of experiments using control point technique are already resume in Table 2. For most experiment using this technique, total station is the additional tool to measure position of targets. Improvement of this technique was done by Bornaz and Rinaudo (2005), by developing software that can be used to automatically doing the registration and georeferencing point cloud in one process. All of the control point coordinate inserted to the TLS system while in scanning can be recognized and then the software will align all of scans part following the overlap area. Experiment by Miri and Varshosaz (2011) proved that accuracy from this method was affected by the distance from scanner to the target and also the direction of the target surface with respect to the laser scanner beams. Increasing distance while scanning may decrease the accuracy.

Data driven technique means the georeferencing process will be done using the existed data such as 3D virtual city model or Digital Surface Model (DSM). This process will be done using Iterative Closest Point (ICP) algorithm. ICP will try to match the surface or geometry formed by point cloud into the existed data. Using this technique, no control point and any other measurement tools are needed. Processing time using this technique is also faster than direct georeferencing (Alba et al. 2007). Though, this technique can only be used when the data is available. Other problem was the availability of DSM data. Matching process using ICP might be hard for building façade since the DSM only have tilt façade. Disadvantage of this technique is that the overlap area should be large enough to cover the surface details (Arayici 2007).

Experiments based on data driven technique were done. Table 3 lists some of georeferencing experiments using this approach. Improvement in accuracy using this approach was made by Schuhmacher and Bohm (2005). LIDAR data and TLS data were integrated; TLS data was georeferenced using ICP. They also used the result of georeferencing using low-cost GPS as a relative orientation to calculate

Fig. 5 Integrated TLS and LIDAR data of Jactin House (Arayici 2007)

Table 3 Results from previous works using data driven technique

Project	Georeferencing technique	Accuracy (m)
Schuhmacher and Bohm (2005)	Integrated with 3D city model	X 0.14 Y 0.08 Z 0.042
	Integrated with LIDAR	X 0.02 Y 0.25 Z 0.08
	Using GPS data as first orientation then integrated with 3D city model	X 0.13 Y 0.08 Z 0.46
	Using GPS data as first orientation then integrated with LIDAR	X 0.05 Y 0.10 Z 0.06
Alba et al. (2007)	Using ICP	Range: 0.007–0.021

the position using ICP. Result of this experiment is shown in Fig. 5. Accuracy resulted from that technique is also included in the Table 3.

4 Conclusion

Georeferencing is an important step in processing TLS data, especially for building model. It generates the possibility to integrate TLS data with other geospatial data. In the process, transforming coordinate system of point cloud or model could be done by various techniques. From the previous works, the highest accuracy was achieved using control point technique. Although, in the thesis by Reshetyuk (2009) shown that the use of direct method also are also possible to get the accuracy below 1 cm.

Control point technique could give high accuracy. However, results from experiments still shown different range of accuracy since the accuracy of georeferencing using this technique is also affected by the distance from scanner to the objects while scanning. Direct technique using GPS offers the efficiency in the field since no control point is needed in the field. Meanwhile, the data-driven approach is offering an innovative technique to do georeferencing with fully or semi automatically with the ICP algorithm. Integration technique offered by Schuhmacher and Bohm (2005) can give improvement in the accuracy of data-driven technique.

Many improvements have been made in order to get the highest accuracy and also automation in the process of georeferencing. As the requirement for building model application, TLS is still a promising measurement tools. Technique in georeferencing for TLS data can be chosen following the requirement. Control point is suitable for application requiring high accuracy. Data driven and Sensor driven are suitable for application constraining in time of data processing.

References

Alba M, Fregonese L, Prandi F, Scaioni M, Valgoi P (2006) Structural monitoring of a large dam by terrestrial laser scanning. In: Proceedings of the ISPRS Commission V Symposium Image Engineering and Vision Metrology, ISPRS, Dresden

Alba M, Giussani A, Roncoroni F, Scaioni M (2007) Review and comparison of techniques for terrestrial 3D-View georeferencing. In: Vettore A, Sheime E (eds) The 5th international symposium on mobile mapping technology. XXXVI-5/C55, ISPRS, Padua

Altuntas C, Yildiz F, Goktape A, Karabork H (2010) A study on measurement of building wall thickness from 3D object model. Int J Phy Sci 5(15):2317–2321

Arayici Y (2007) An approach for real world data modelling with the 3D terrestrial laser scanner for built environment. Autom Constr 16(6):816–829

Aydar U, Akyol O, Duran Z (2011) A low-cost laser scanning system design. In: XXIIIth International CIPA Symposium, Prague

Baz I, Kersten T, Buyuksalih G, Jacobsen K (2008) Documentation of Istanbul historical peninsula by static and mobile terrestrial laser scanning. In: Jun PC (ed) XXIst ISPRS Congress. XXXVII, Beijing, ISPRS

Bohm J (2005) Terrestrial laser scanning: a suplementary approach for 3D documentation and animation. Photogram Week 05: 263–271 (Wichmann Verlag, Heidelberg)

Bornaz L, Rinaudo F (2005) Terrestrial laser scanner data processing. In: XXth ISPRS Congress, Istanbul, ISPRS, 12–23 July 2004, pp 514–520

Boulaassal H, Landes T, Grussenmeyer P (2011) Reconstruction of 3D vector models of building by combination of ALS, TLS, and VLS data. In: Remondino F, Hakim SE (eds) ISPRS workshop 3D-Arch 2011. XXXVIII-5/W16. Trento, ISPRS

Doellner J, Bucholz H (2005) expressive virtual 3D city models. XXII international cartographic conference (ICC2005), A Coruda, Global Congress

Dursun S, Sagir D, Buyuksalih G, Buhur S, Kersten T, Jacubsen K (2008) 3D city modelling of Istanbul historic peninsula by combintaion of aerial image and terrestrial laser scanning data. The international archives of the photogrammetry, remote sensing, and spatial information sciences, XXXVII (Part B7)

Gonzales-Aguilera D, Gomez-Lahoz J, Munoz-Nieto A, Herrero-Pascual J (2008). Monitoring the health of an emblematic monument from terrestrial laser scanner. Nondestr Test Eval 301–315

Guidi G, Remondino F, Russo M, Menna F, Rizzi A (2008) 3D modeling of large and complex site using multi-sensor integration and multi-resolution data. In: Ashley M, Hermon S, Proenca A, Reodriguez-Echavarria K (eds) The 9th International Symposium on Virtual Reality, Archeology and Cultural Heritage. The Eurographics Association, Braga, Portugal

Harrap R, Lato M (2010) An overview of LIDAR: collection to application. Norway

Karabork H, Altuntas C, Tusat E (2012) Combining of scene measurements by laser scanner and GPS combination. FIG Working Week 2012. Rome, Italy

Lee IS, Lee J, Ge L (2009) Comparison of terrestrial laser scanner with digital aerial photogrammetry for extracting ridges in the rice paddies. Surv Rev 41(313):253–267

Miri M, Varshosaz M (2011) Evaluating parameters affecting the georeferencing accuracy of terrestrial laser scanners. In: Remondino F, El-Hakim S (eds) ISPRS Workshop 3D-ARCH 2011 3D virtual reconstruction and visualization of complex architectures. XXXVIII-5/W16, Trento, ISPRS

Paffenholz J-A, Kutterer H (2008) Direct georeferencing of static terrestrial laser scanners. FIG Working Week 2008, development in scanner and sensor technologies, Stockholm

Reshetyuk Y (2009) Self-calibration and direct georeferencing in terrestrial laser scanning. Royal Institute of Technology, Stockholm

Scaioni M (2005) Direct georeferencing of TLS in surveying of complex sites. In: Hakim SE, Remondino F, Gonzo L (eds) 3D Arch: virtual reconstruction and visualization of complex architectures. XXXVI-5/W17, Mestre-Venice, ISPRS

Schuhmacher S, Bohm J (2005) Georeferencing of terrestrial laser scanner data for applications in architectural modeling. In: Sabry El-Hakim FR (ed) 3D Arch: virtual reconstruction and visualization of complex architectures. XXXVI-5/W17, Venice, ISPRS Archives

Sester M (2007) 3D visualizationa and generalization. In: Fritsch D (ed) Photogrammetrische woche. Hannover, IFP Stuttgart

Shi R, Zhu G, Zhu L, Wang R, Mu Y (2009) Digital construction for spatial documentation of Beijing old city. 22nd CIPA SYmposium, 11–15 Oct, Kyoto, CIPA

Strzalka A, Bogdahn J, Coors V, Eicker U (2011) 3D city modelling for urban scale heating energy demand forecasting. HVAC\&R Research, 17(4):526–539

Susaki J (2009) Modeling of building from terrestrial laser scanning considering topology for occlusion. 22nd CIPA Symposium, 11–15 Oct, Kyoto, CIPA

Vela E, Babic L, Dapo A, Kordic B, Pribiceviv B, Medak D (2010) Terrestrial laser scanning for the digital preservation of a croatian historical village Dobranje. The XXIV FIG International Congress 2010, Sidney, FIG

Voegtle T, Steinle E (2000) 3D modelling of buildings using laser scanning and spectral information. International Archives of Photogrammetry and Remote Sensing. XXXIII, Amsterdam, ISPRS, pp. 927–934

Vozikis G, Haring A, Vozikis E, Kraus K (2004) Laser scanning: a new method for recording and documentation in archeology. FIG Working Week 2004 —Archeological Surveys. WSA1 Recording Methods, Athens, FIG (Austria/Greece)

Yang B, Li Q, Li D (2002) Building model creating and storing in 3D urban GIS. XIX ISPRS Congress Comission I-VIII. XXXIII Part 4. Amsterdam, ISPRS

Yildiz H, Gumusay U (2011) 3D modeling of the cukursaray (The Hollow Palace), Istanbul—Turkey and its application for campus information system. XXIIIth International CIPA Symposium Sep 12–16, Prague, Czech Republic, CIPA

CityGML for Architectural Heritage

Erik Costamagna and Antonia Spanò

Abstract Standardization in spatial data and Cultural Heritage management is one of the main issues in this field of study and the efforts in this direction are twofold, data models and technological development. Regarding the first aspect some preliminary steps towards definition of common vocabularies and conceptual schemas for semantic classification will be analysed and discussed in order to define the open issues and the development strategies. On the side of the modelling languages and data manipulation tools the recent standard definitions and the research efforts are starting to provide a set of application tools for the semantic and spatial data modelling and information retrieving and manipulation. An application of these notions and tools will be performed on an Architectural Heritage case-study and the preliminary results will be exposed.

Keywords CityGML · Architectural Heritage · XML-based data formats · XML schema extension · Spatial data management

1 Introduction

This study will focus on spatial data modelling and management, with particular attention to open standards and modelling languages. The aim of the research is to try to define path for a spatial data management through native XML databases and the development of data models for specific fields of application like Cultural Heritage (CH) documentation. In order to achieve these objectives a brief

E. Costamagna (✉) · A. Spanò
Politecnico di Torino, viale Mattioli 39 10125 Torino, Italy
e-mail: erik.costamagna@polito.it

A. Spanò
e-mail: antonia.spano@polito.it

A. Abdul Rahman et al. (eds.), *Developments in Multidimensional Spatial Data Models*, Lecture Notes in Geoinformation and Cartography,
DOI: 10.1007/978-3-642-36379-5_14,

overview about the standards and languages in spatial information and CH documentation used in the researches is summarized.

1.1 Modelling Languages

In this brief overview we will focus on the logical formalization of conceptual schemas for the implementation of spatial information in XML databases. The consistency and completeness of the data translation through different levels of data definition is one of the major issues in the Database Management Systems (DBMS) field. The XML technology offers efficient tools for the translation of conceptual models in logical schemas and object instances serialization, implementing some of the object-oriented notions. The Unified Modelling Language (UML) (OMG UMLv2.4 2011) is a standard modelling language in the field of object-oriented software engineering. The basic notions and tools of this approach have been applied also to databases (Atkinson et al. 1990). Today it is one of the most widespread tools for conceptual modelling and it is the common grammar used by several standardization organizations, including International Standard Organization (ISO) and Open Geospatial Consortium (OGC). Its major features like object identity, encapsulation, inheritance and polymorphism represent a step forward towards providing database tools for representing complex data structures like spatial data (Egenhofer and Franck 1992; Peckham and Maryanski 1988).

1.1.1 XML Technology

XML metalanguage (2008), a World Wide Web Consortium (W3C) standard, is one of the mainstay tools of the semantic web, widespread in a large set of applications. It allows information structuring and exchanging over the net. The development of this technology, initially oriented to the representation of textual information (document-oriented approach), led to an integration with the DBMS featured in a different approach to information retrieving and manipulation from the widespread relational databases. These DBMS are known as NoSQL databases as the complex structure of the document-oriented approach led to the development of specific query languages, like XQuery. XML technology's strongest points are the flexibility of the data model, which allow several extensions for specific fields of application, and the structure for information description and indexing. These features led to the development of several modelling languages based upon this standard, including the spatial data ones. Some examples of XML based languages for geometry data modelling are Industry Foundation Classes (IFC) (2003) and Extensible 3D (X3D) (2008). IFC is the emerging standard for the Architectural Engineering and Construction (AEC) sector. It is the data model and a data storage format for the Building Information Modeling (BIM) paradigm which bring into the CAD world some of the database tools and concepts. Several studies has been made upon the integration with IFC data model and City

Geographic Markup Language (CityGML) (Zlatanova and Isikdag 2009). These efforts are a remarkable step forward for filling the gap between the CAD systems and GIS.

The structure of an XML document is based on a schema which is used to provide the vocabulary to decode the information and a tool to validate the data structure. This schema represents the essential means to provide the portability of data among different development tools. Each XML document can be validated against a schema which must be declared and available for parsing. The XML schema format recommended by the W3C is the XML Schema Definition Language (XSD) which provides a definition for the basic data structure (XSD1 2004) and datatypes (XSD2 2004). This model allows the representation of complex data structures by means of the extension mechanism which is an implementation of the object-oriented inheritance notion and UML constructs. UML constructs are implemented by means of elements, representing classes, simple and complex types, representing attributes and relationships. The object-oriented approach provides instruments for representing complex data structures like spatial data. This feature can be implemented in the XML schema without any loss of semantic value and can also be automatically derived through reverse-engineering techniques.

XPath

XPath (2010) is the language designed to navigate XML documents. Since the basic unit of XML data is the document root, it is necessary to provide a specific tool for the node selection. Based upon the knowledge of declared XML schemas, with this tool it is possible to traverse the nodes of the XML document structure allowing the interrogation of specific attributes with the XQuery operators.

XQuery

XQuery (2010) is a language developed to perform query instances against XML documents. The FLWOR (*For Let Where Order Return*) logical operator can be considered the counterpart of the Structured Query Language (SQL) in XML environment. An SQL statement can be easily logically formalized and its instance can be represented by one or more tuples of a table in a relational database. On the other side a FLWOR query instance is more complex since also the data structure of XML documents is more complex. The *For* clause of this operator specifies the sequence of XML nodes, the *Let* clause specifies the association of the variable with the node while the *Where* clause is the core part of the operator representing the query condition. The *Order* clause allows the nodes ordering against which the *Where* clause is instanced and the *Return* clause gives back the results for each node.

XML Data Binding

This extension of the XML language functions is a tool for the translation of UML schemas into XML ones (*marshalling*) and vice-versa (*unmarshalling*). This process can be performed automatically without any loss of information due to the implementation of specific object-oriented functionalities in the XML technology.

1.2 XML Based Standards in Geographical Information

In the survey of normative international institutions, Open Geospatial Consortium (OGC) is the major source of open models and languages for encoding spatial information. OGC standards, like GML, have been adopted by most of the mapping agencies and they are starting to be supported by GIS software tools, which are the prerequisites for their spreading among the specialists in this field. Some of the widespread OGC standards are Keyhole Markup Language (KML) (2008), GML and CityGML. All of them are based on the XML language and in this study we will focus on the last two as they fully support the third dimension.

1.2.1 GML

The OGC Geography Markup Language (GML) standard (2007) deals with the definition, representation, storage and exchange of Geographic Information. Developed since 2000 by the OGC, the GML standard has meanwhile reached version 3.3. The conceptual model has been logically implemented with the XML technology tools. The language allows for the modelling of the majority of 2D and 3D vector representations and also for raster representations. The conceptual basis of the model combined with the completeness and flexibility of the XML technology makes this language a common grammar to build upon extensions for specific domains like urban data management.

1.2.2 CityGML

The OGC City Geography Markup Language (CityGML) is an OGC standard (2012) developed by the Special Interest Group 3D (SIG3D) of the Initiative Geodata Infrastructure Germany (GDI-DE) in cooperation with the OGC. The goal of this specification was to define a data model for describing thematic data in urban mapping by means of GML. Indeed, CityGML is an application schema of the GML language. The CityGML schema is modular, composed by a mixed geometrical-topological module and several thematic modules corresponding to the urban map layers. Compared to the GML language, and in addition to the semantic specialization, CityGML has developed multiscale and multiview support. The CityGML approach to multiscale modelling consists in a feature discretization based on the relationship between the geometric and thematic component. For each Level of Detail (LoD) there is a specification for the generalization level, the absolute tolerance and the included thematic classes. The model allows to associate several instances of a geometric class to each instance of a thematic class through defining a specific relation for each LoD. In addition, the multiview support associates a set of visualization settings to each instance of a thematic class, including photo-realistic rendering of the surface of geometric models. A specific module (*Appearance*) allows to manage the visualisation independently of the geometric and thematic class implementation, thus each instance of a geometric-thematic class can share different visualisation settings (*themes*).

1.3 Availability and Current Standard Gaps for Cultural Heritage Documentation

In broad terms, the development of a language, of an International grammar for managing the documentation pertaining CH is a topic which has become urgently addressed in the last few years. The topic is manifest especially in response to the requirements made by the scientific communities and preservation bodies of those countries which made a contribution in highlighting and spreading the knowledge. The CH sector greatly benefits from universal sharing of techniques and procedures, resulting in the increase of the value of the heritage itself. Concerning CH, shared decision are more difficult to achieve with respect to other sectors whose internal structures are more evened out, given the fact that this sector is more multi-disciplinary in nature, involving experts coming from several fields of study (archaeology, history, architecture, arts, but also technology, information systems, geomatics, geo/topo/cartographic sciences), which should cooperate and above all should refer to a single and common international institution, presiding over quality control of the work performed and assessment of International compatibility. A non-negligible problem lies in the vast classification of the heritage relative to museum collections, libraries and archives. A starting point for standardization in the CH sector, addressing these issues, is ISO 21127 (2006). The regulation framework in the CH sector is further complicated by the presence of several organizations, institutions, associations which produce and process technical documents lacking, however, all guarantees concerning a consensual acceptance of procedures and of conformity with other standards. Considering CH cataloguing, numerous international organizations are moving in this direction, establishing guidelines for CH documentation.

On the international scale, the debate concerning the definition of the standards for Cultural Heritage has found full expression in the International Committee for Documentation (CIDOC), a committee of the International Council of Museums (ICOM), which supplied a model for documenting museum objects, corresponding to ISO 21127. In the United Kingdom, the Royal Commission on the Historical Monuments of England, together with the European Council, proposed the Core Data Index to Historic Buildings and Monuments of Architectural Heritage (Thornes and Bold 1998). The more comprehensive standard meta-data models are Categories for the Description of Works Arts (CDWA), proposed by the Getty Information Institute of Los Angeles which supply detailed instructions for the scientific description of the objects. We can find a similar scheme in the Virtual Resources Association (VRA) Core Categories, mainly used for archives of visual material. CDWA and VRA Core are recent standards, which, thanks to the increased advancement of digital resources, play a key role in the improvement of descriptions and access to cultural resources, and their employment aids in transfer ring data to the new systems. By way of guarantee of interoperability, both CDWA and VRA use the same CCO guidelines for describing contents and offer XML encoding for sharing and exchanging data.

According to our current experience in handling Architectural Heritage 3D metric data in the wider framework of urban spatial data management (Sect. 2), despite the availability of archiving standards, it is necessary to define a common vocabulary for classifying the geometric and thematic content of constitutive parts of buildings. There are consolidated methods and guidelines, or cataloguing rules which have been developed for other employments, often of a thematic and sector nature. Their aim is to guide the choice of parts of buildings and structural and architectural elements whose specific features identify the piece of heritage intended for classification. These are represented by card A of the Central Institute for Catalogue and Documentation—Istituto Centrale per il Catalogo e la Documentazione (ICCD), of the Italian Ministry for Cultural Heritage and Activities, by guidelines adopted by the scientific community of preservers in order to establish the maintenance or restoration activities or by other extremely specific guidelines. Examples are the ones addressing the risk assessment and reduction of the preserved Cultural Heritage[1] (2011), with reference to the technical regulations for masonry buildings, which also identifies levels of hazard and risk thresholds by analysing typological and structural elements of the buildings. Identification classes of the buildings' constituent parts are assigned entry texts which specify typological categories, also in reference with building/typological systems of a local nature (ICCDpre 1992). Table 1 shows these classes identifying the building components.

[1] "Assessment and reduction of seismic risk of the Cultural Heritage with reference to Technical Regulation for Buildings as indicated in the Decree of the Ministry of infrastructure and transport dated 14th January 2008". Directive by the President of the Council, Official Journal No. 47 dated 26th February 2011—Ordinary supplement No.54.

Table 1 Identification classes of the components of buildings according to card A of ICCD

IS	STRUCTURAL LAYOUT
PN	PLAN
FN	FOUNDATIONS
SV	VERTICAL STRUCTURES
SO	HORIZONTAL STRUCTURES
CP	COVERINGS
SC	STAIRS
PV	FLOORS AND FLOORINGS
DE	DECORATIVE ELEMENTS

2 CityGML Model Implementation

The model implementation within the CityGML data structure is the first step of the XML database building and an opportunity to test the features of the OGC data model architecture. The reasons of this choice, apart from the features of XML language on which it is based, are the targeted urban data classes definition and multiscale and multiview support. These features, which are specific implementations of the CityGML standard from the base structure of the GML model, provide the specific tools for representing the complexity of spatial data in urban and architectural context.

2.1 The Base Dataset

The Valentino Castle is a seventeenth century mansion located in the urban fabric of Torino and is one of the Residences of the Royal House of Savoia, included in the UNESCO World Heritage List since 1997. The building stands on the west riverbank of the Po river, in the southern part of the Roman city. Some of the features of this building, like the spatial complexity, the layering of the constructive phases, its relationship with the urban fabric and its present function as university, illustrate the main issues of the management of Architectural Heritage in an urban context. Its spatial extension is the result of the composition of its parts: the pavilion towers, the main part, the galleries and the court of honour precisely define the typology of the palace, a clear derivation from the French style. Another distinguishing mark is the way in which it connects with the urban fabric, the river, the surrounding park and the hilly landscape over the eastern view of the castle. For these reasons it can be taken as a typical example of the issues in documenting Architectural Heritage in an urban context. This building was involved in a series of survey campaigns between 2007 and 2009. They were part of the preliminary step for the restoration of the building façades which has already been partially realized. These survey campaigns have been carried out by our group, specialized in image-based survey methods and other geomatics techniques applying to CH. The survey involved the exterior façades, the courtyard and part of the interiors and the survey products, consisting of high detailed georeferenced raster and vector representations, were obtained from photogrammetrical, topographic, Terrestrial Laser Scanner (TLS) and Global Navigation Satellite System (GNSS) data capture.

Even if there are no standards for metric documentation, and their adoption is therefore not required, in Italy, the metric survey of the Valentino Castle and its products have been carried out with good compliance with the English Metric Survey, a normalisation standard for metric surveys developed by the Historic Buildings and Monuments Commission for England (Bryan et al. 2009). Despite the existence of this gap, the preservation bodies tend to require that the survey be conducted through geomatic techniques, and that the products of the survey be manageable in organised metric archives accessible and up-to-date. The methods which define Geomatics guarantee a series of invariable conditions characterising the survey model:

- it's feasible that a general 3D spatial archive can be generated, where processed and raw data and even model results or other kind of representation can coexist; the updating process is expedited and cultural item documentation can be developed as a continuously evolving process;
- 3D data archives present different resolution and accuracy according to documentation plans because different specialized analyses have to be allowed, diverse behaviours and phenomena of the cultural item have to be investigated in different in-depth levels;
- large amounts of data can be collected and reconstructed 3D models are textured by restrained images, the integration of vector and texture data is well estimated in the dissemination phase; in case of buildings subject to conservation and restoration plan, survey methods involving surface description or image texturing are essential to highlight problems distressing the whole building, such as plaster degradation, subsiding or other anomalies like troubling structures, moistness and other environmental causes of damage.

This was the base dataset, corresponding to the most widespread typologies of metric documentation in the CH domain, from which 3D models at different LODs were built. The qualitative feature of raster representations like rectified images and ortho-photos, which are essential tools to understand the cultural value of the object, were combined with the 3D vector models. Figures 1, 2 and 3 show some examples of the drawings produced that were used to build the restoration project.

2.2 Multiscale Support

Multiscale modelling is an approach to the understanding of spatial information that requires a discretization of geometric features in predefined levels of detail (LODs) based on their semantic values. The morphological, functional and

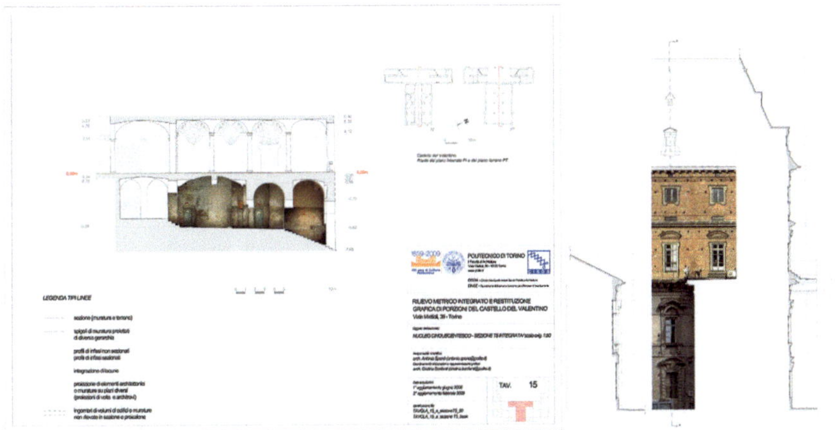

Fig. 1 *Cross sections* showing results of integrated 3D survey with rectified image mosaic (drawing: C. Bonfanti)

constructive features are the main categories for the comprehension of architectural objects. The Architectural Heritage consists of a complex structure of relationships between geometric and thematic information, which are evident when switching from the urban scale to higher ones. A multidisciplinary approach to building feature classification must take into account both the coded languages in the architecture criticism and the data management issues (Apollonio et al. 2010).

The CityGML model supports 5 LODs each one of them consisting of specific relations between thematic and geometric classes. The LOD definition is logically formalized in such a way as to be independent from the application tools. Even if it is based on these characteristics of independence, the definition of the levels of detail upon which the representation of the geometric objects is based, suffers necessarily from what is already consolidated in the semantic definition and in the criteria of generalisation of regional and urban cartography at different scales (between 10,000 and 1,000). Even if a considerable amount of new spatial information will enrich the middle scale between the urban and architectural scales (between 1,000 and 100), it is indubitable that currently available CityGML specification can be considered effective for describing and

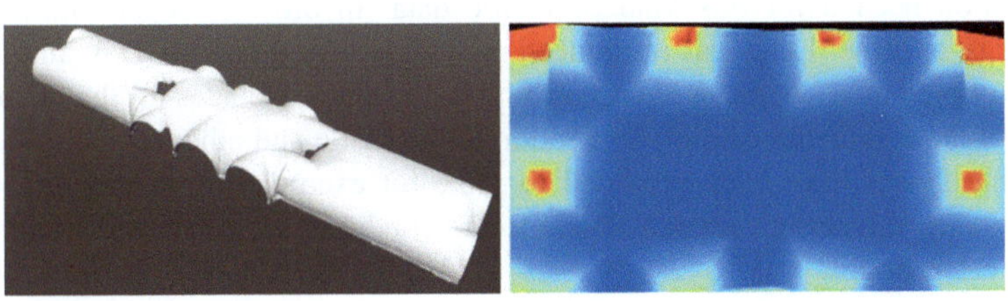

Fig. 2 The 3D laser model and the DEM of the vault of first level loggia of the castle

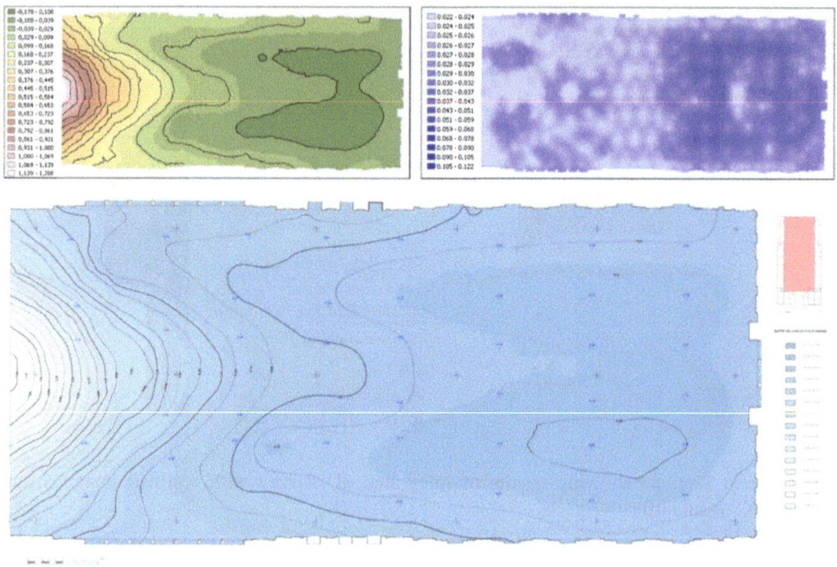

Fig. 3 On the *top* the DEM generated using kriging interpolation with overlaying of the contour lines and the variogram, following the geometric drawing of the flooring as a function of the points distribution. On the *bottom* the integrated representation of the courtyard surface supporting the ordinary maintenance of drainage system

managing of the urban fabric but not completely so for the pieces of Architectural Heritage located therein.

Furthermore, it is necessary to highlight that the semantic and geometric definition contained in the CityGML standard has taken into consideration the current methods for acquisition of spatial information at the urban scale, therefore meaning those systems which provide for integration of Digital Surface Model (DSM) acquired by low flying Light Detection And Ranging (LiDAR) passes, with those deriving from the so-called Mobile Mapping Systems (MMS). The latter are surveying systems which combine surveying tools, by means of a 2D laser scanner and images acquired by a certain number of cameras able to acquire partially superimposed scenes, or cameras with spherical lenses able to take in 360° panorama shots. Given the fact that the CityGML approach is eminently urban, the standard's building classification doesn't fulfill the semantic and geometric specialization needed for this kind of metric data. In order to repair these shortcomings, a further LOD was introduced, at a higher scale than LOD 4, and which, for the time being, we have identified as LOD*, which includes all the decorative parts of the eighteenth century decoration of the castle and other specialisations of existing classes have been provided for, like for example the duplication of the Appearance theme, in order to illustrate the condition of the façades before and after restoration.

2.3 Multiview Support

Multiview modelling defines the capacity of a system to describe the surface aspect of geometric objects in different ways corresponding to specific point of views or domain of interest through a parameterization of these features. Some of them are intrinsic to the material feature like colour, transparency and smoothness while others like illumination, reflection and refraction are environment-related. The management of multiview modelling is a typical feature of CAD systems, including tools for material and environmental parameterization and photo-realistic rendering. These are some of the CAD functions added to GIS for the management of 3D architectural models (van Oosterom et al. 2005). Unlike CAD systems the CityGML model provides different classes for materials and texture modelling. The materials are defined through X3D specification tools of Web3D Consortium, while the texture classes allow the management of different kinds of representations like homographic and orthographic projections, texture parameterization and georeferenced images. The raster representations are of great importance in CH metric documentation since they are used to highlight qualitative features of objects like the preservation state and the construction material. They are also useful to record the state before and after a restoration intervention in order to manage the building's lifecycle. With this purpose new photograms of the southern façade were taken and the rectified images of the previous and subsequent state were implemented in different *Appearance* themes (Costamagna and Spanó 2012). This implementation is shown in Fig. 4.

Fig. 4 The two *Appearance* themes (LOD2-4) implemented: before restoration (*top*) and after (*bottom*)

2.4 Thematic Model

The *Building Module* is the most complex scheme in the CityGML specification and the richest in semantic and geometric definition including the interiors and exteriors of a building. The first semantic specialization is represented by the *BuildingPart* class associated with the *Building* class by means of the *consistOf BuildingPart* aggregation relation allowing recursive aggregation between them. The *Building* and *BuildingPart* class features are defined through several attributes, representative of the main characteristics of the structure. The plan structure of the Valentino Castle, derived by the French royal palace style, consisting in a series of building parts, differently shaped, surrounding a courtyard is the material representation of the absolute power of monarchy and this feature is the most representative of the whole building. To formalize the relationships between the building parts and the whole complex the *BuildingPart* to *Building* association was modelled as a compositive relation. So all the parts of the castle were implemented at LOD1 as instances of the *BuildingPart* class while the *Building* instance is defined by the aggregation of its parts. Starting form LOD2 the specialization of the building parts allows to define the different elements of the building envelope such as the *GroundSurface*, *RoofSurface* and *WallSurface* classes, associated with the *Building* class through the *boundedBy* relation. The LOD3 allows to model the openings of the façade while the LOD4 allows to represent the interiors. In spite of their comprehensiveness in representing architectural features, CityGML thematic classes does not fulfill the semantic specialization of the historical buildings. The decorative elements which are evident from the LOD3 should be modelled with proper class specializations of the *BoundarySurface* classes. Figure 5 shows the schema of the geometric-semantic generalization of the implemented CityGML classes.

3 Cultural Heritage Application Domain Extension

Here we will showcase a proposal for an extension of the CityGML schema in order to provide a outline for the conceptual and logical modelling of an Architectural Heritage feature. This Application Domain Extension (ADE) draft will focus on modelling a typical Architectural Heritage feature that can be useful in planning restoration works and conservation state management. The aim is to use this extension for performing diagnostics identifying current and previous phenomena, whose comparison is of capital importance for planning proper conservation. The main issues of the management of a CH archive will be taken into account in order to integrate the rich semantic features of this kind of metric data in urban cartographic databases. Architectural Heritage (UE1985 1985) is a part of the larger and more complex set of CH. A complete analysis of the CH features exceeds the scope of the study that will focus on the development of specific classes and relationships for the CityGML *Building* module extension. The

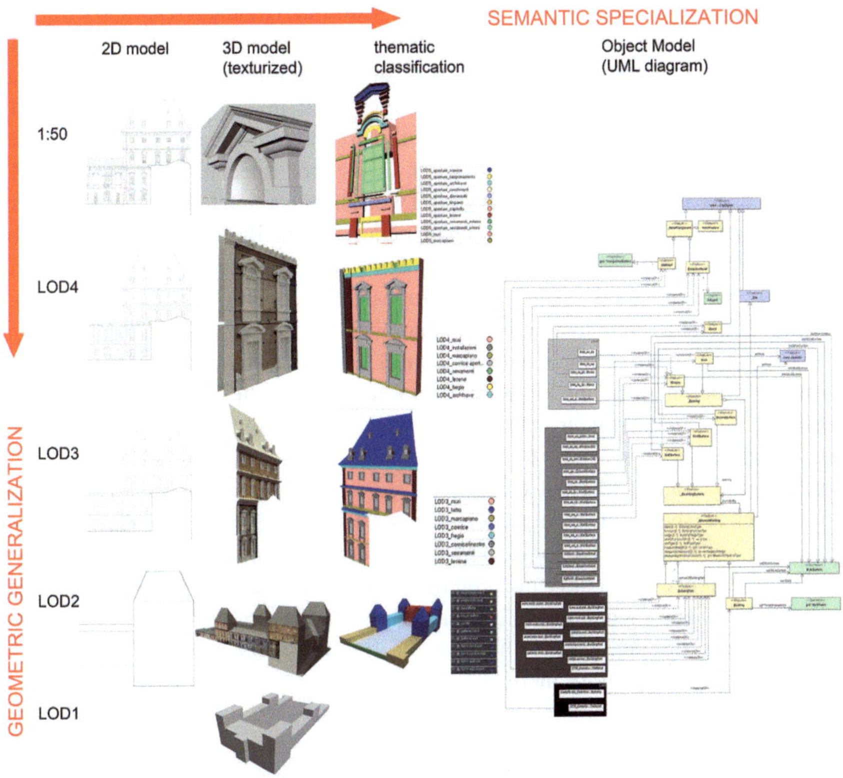

Fig. 5 Workflow of the semantic and geometric modelling process

relationships between Architectural Heritage features and the other CH will be considered only with the purpose to define a framework to integrate other ADEs in the CityGML module for urban data management. The *Building* module will be extended by means of two levels of in-depth analysis: in a first step new attributes for the *Building* class and its derivations will be created in order to model the common features that define a building as Cultural Heritage. The second step, a closer look of the *Building* module classes, will involve the modelling of a common feature of Architectural Heritage: the building façades deterioration phenomena. All these features will be modelled using the UML and XML Schema tools (Apollonio et al. 2010).

3.1 UML Model

As outlined previously the features of Architectural Heritage are a part of a larger set as the CityGML model thematic classes are derived by the GML superclass *Feature* which represent a general container for all the geographical thematic classes. The modelling of a subset of the Cultural Heritage features will have classes, relationships and attributes in common with both the GML model and its specialization CityGML. In order to allow an integration in the model of different kinds of CH, some preliminary issues were taken into account. All CityGML thematic classes, as they are designed for the urban data management, are derived from the abstract class *CityObject* which is the first specialization of the GML *Feature* class introduced in the model. Two attributes are featured in this class: *creationDate* and *terminationDate* which clearly define the temporal classification of urban data. The GML model on the other side provides a specific module for modelling the temporal phenomena composed by specific classes and relationships. It is quite evident to every technician who has approached Cultural Heritage that the *CityObject* attributes may be a strict limitation if we enlarge our point of view, embracing different kinds of Cultural Heritage like archaeological sites. Setting the Architectural Heritage in a wider context it might be useful to allow the implementation of specific data features without constraining the model with predefined attributes like the *CityObject* ones. Following these principles a specialization of the GML *Feature* class has been introduced: *ch::_CulturalHeritageFeature*. This can be considered the superclass from which all the Cultural Heritage features which have in common the geographical information content can be modelled. The model specialization follows the layered structure of the CityGML model introducing a specific class for the hand-made cultural features as the *_Site* class of the *Building* module: *ch::_CHSite*.

Once we laid the foundations of the CH ADE model specific classes, relationships and attributes of the architectural features can be introduced. First of all in order to set the basic tool for indexing and retrieving Architectural Heritage in urban spatial databases we need to implement the attributes of the *Building* class. These implementations are: *isCH, isCHAuth, CHAuthName* and *CHAuthCode*. These attributes will be later defined using the XML and GML datatypes. The specific CH features to be modelled involve a common kind of CH metric documentation which is the preliminary step for any kind of restoration intervention and conservation state monitoring: the deterioration state documentation. These kinds of data are usually processed in order to compute the amount of the interventions needed and the relative costs. Translating these notions into the UML modelling constructs, an abstract class, specialized from the *ch::_Site* one, will be introduced: *ch::_AbstractDeterioration*. From this class the *ch::_BoundarySurfaceDeterioration* class is created. This will be the base for the specialization of specific classes for different kinds of deterioration such as *ch::WallSurfaceDeterioration* and *ch::OpeningFrameDeterioration*. At this level the aggregation relationship between the *_BoundarySurface* class and the *gml::MultiSurface* geometry class is modeled. Figure 6 shows the UML schema of the CityGML *Building* module extension.

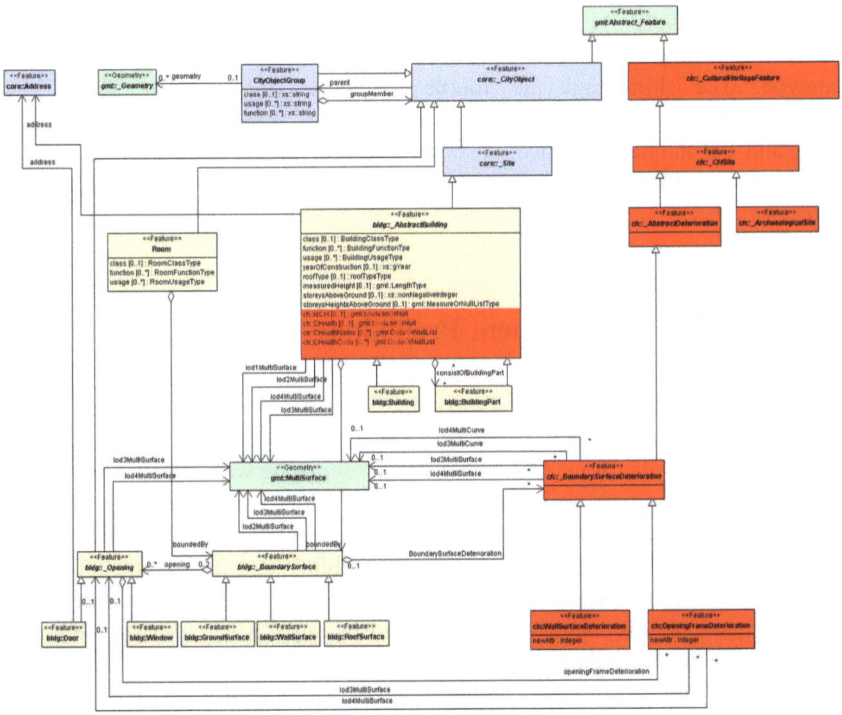

Fig. 6 UML schema of the *Building* extension

3.2 XML Schema

The translation of the UML conceptual model into the XML Schema is an easy data restructuring process using the extension tools of the XML language. Each class has been broken down into a complex type and its related element. Consequently the specialization associations are defined through the *base* and *derivedBy* attributes for the complex type and the *substituitionGroup* for the element. The complex type *BoundarySurfaceDeteriorationPropertyType* is created in order to model the aggregation relationship between the *ch::_BoundarySurfaceDeterioration* and the *_BoundarySurface* class. At the schema definition level it is possible to define the datatypes used to model specific attributes such as those introduced for the *Building* class implementation. These attributes are designed in order to allow the integration of Architectural Heritage data in a common spatial database archive. The first attribute introduced is *isCH* with [0...1] cardinality and a *gml::booleanOrNull* datatype. This attribute defines the CH value with a particular kind of boolean value (*nillable*) that can be nullified with an empty value, eventually specified through a coded list (*nillableReasonList*). The possibility to leave empty the value may be useful in case it is not possible to assign this property to the object at a particular moment. Following the same schema a second attribute is defined: *isCHAuth* [0...1] targeted to the definition of the authority statement about the cultural value which will also be expressed with the *gml::booleanOrNull* datatype. The third and the fourth attributes, *CHAuthName* [0...*] and *CHAuthCode* [0...*], define the name of the authority which is stating the cultural value and the unique identification code assigned. Figure 7 shows the complex types *_CulturalHeritageFeature*, *_AbstractDeterioration*, *_BoundarySurface Deterioration*.

4 Spatial Data Management Preliminary Issues

The application of XML technology to database management is quite a recent step forward in the development of portable tools for the data manipulation among different software platforms. One of the aims of semantic web technologies, which XML language is a part of, is the provision of languages, ontologies and formal schemas. These tools allow data reading and manipulating across different software platforms, avoiding the lack of compatibility due to application based solutions for information structuring and data retrieval. Extensions of data model and data manipulation languages can be read by any XML parser providing the necessary schemas and Application Programming Interfaces (APIs). XQuery and XPath tools allow to formulate complex queries against semantic content of XML documents. Figure 8 shows a complex thematic query against a *BuildingPart* attribute.

One of the frontiers in XML technology development is the implementation of the spatial support for XQuery and XPath languages. A standard proposal for a geospatial API interface for XPath 2.0 has been developed by the EXPath initiative (2010). The EXPath Geo Module is an implementation-independent API designed to work with XPath 2.0 based languages, like XQuery, providing specific functions

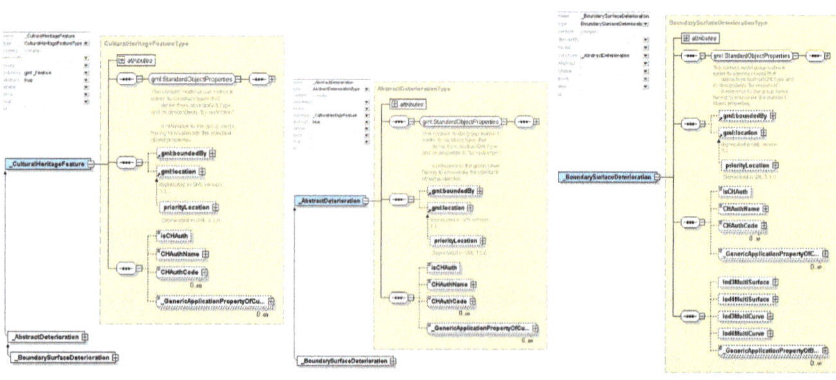

Fig. 7 XML schema implementation. From *left* to *right* the complex types: *_CulturalHeritage-Feature, _AbstractDeterioration and _BoundarySurfaceDeterioration*

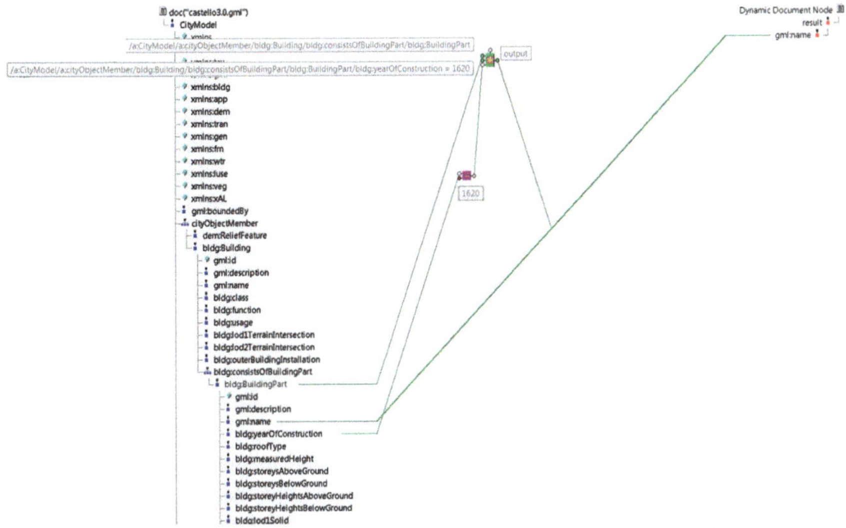

Fig. 8 Visual representation of the FLWOR operator: the Where clause in the magenta box defines the value (1620) of the *yearOfConstruction* attribute, while the others in the *green box* For (*top*) *small* Let (*bottom*) and Return (*right*) define the node list to be analysed, the variable association and the result to be shown

for the GML geometry classes, including spatial analysis operators. Most of the GML Spatial Model geometry primitives and aggregates are supported as the predicates of the Dimensional Extended nine-Intersection Model (DE-9IM) developed by Clementini and Egenhofer (1990, 1993). The aim of this tool is to provide a basic vocabulary and logical operators that can be implemented by specific APIs in any XQuery processor and XML parser. Instead of application driven implementations of the XPath and XQuery language this kind of approach allows the portability of query instances across different software platforms. At the moment there are two software implementations of the EXPath Geo Module which follow two different implementation approaches. The first one is the spatial module of the eXist database suite, in which spatial analysis functions of Geo Module are implemented with an in-the-box approach, thus modelling the functions within the application environment. The second one, the Zorba XQuery processor, is an off-the-shelf open-source implementation that allows assembling this API in different software platforms.[2] The Zorba XQuery processor has been tested within the Eclipse modelling tool, an open-source software platform that allows the integration of different packages for the creation of targeted software solutions (Fig. 9).

---

[2] see zorba-xquery.org, exist-db.org and eclipse.org for further information.

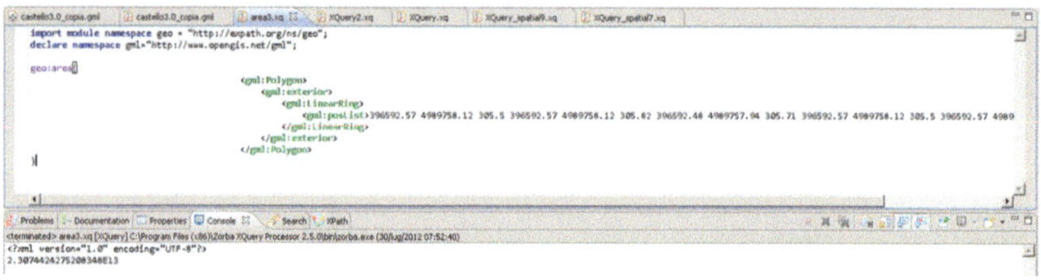

Fig. 9 Result of a simple spatial query on a GML data structure with the Zorba XQuery processor within the Eclipse environment

5 Discussion

In this study we tried to tackle a complex set of spatial data typologies in relation to the data modelling and manipulation tools. The standards, languages and application tools chosen, are designed and developed in order to share information over different cultural context and technology tools. This feature is a common need both in the spatial data management and its subset of Cultural Heritage metric documentation. The XML based languages like GML are spreading over different application fields while targeted management application tools are starting to be developed and exchanged allowing their use across different software platforms. The results shown demonstrate that these tools are potentially capable of handling the semantic structured content of geographical information performing the standard spatial analysis functions of GIS. The XML Schema extension is a little sample of the flexibility and strength of the data modelling tools of this technology while the languages and application interfaces to manage also the spatial objects and operators are under development. The application of this tools at the Cultural Heritage features subset is perhaps one of the most difficult challenges since this kind of features have an high degree of complexity tied with uniqueness traits associated to its specializations. These traits, more than the complexity, are the hardest to match with the standardization processes. The consciousness that standards enable to improve the value of cultural assets, adding to the capability of protection and promotion, can aid the development of shareable standards. Good results have been overtaken in cataloguing data archiving, and the availability of tools will certainly support standards adoption.

References

Apollonio FI, Gaiani M, Manferdini AM (2010) Modellazione sematica metodi a multirisoluzione. In: Modelli digitali 3D in archeologia: il caso di Pompei, chap. 8 Modellazione semantica metodi a multirisoluzione, pp 234–267 (Edizioni della Normale).

Atkinson M, Bancilhon F, De Witt D, Dittrich K, Maier D, Zdonik S (1990) The object-oriented database system manifesto. In: Jagadish HV, Garcia-Molina H (eds) Proceedings of the ACM SIGMOD international conference on management of data. ACM Press, ACM - SIGMOD

Bryan P, Blake B, Jon B (2009) Metric Survey Specifcations for Cultural Heritage. English Heritage. http://www.english-heritage.org.uk/publications/metric-survey% -specification/

Clementini E, Di Felice P, van Oosterom P (1993) A small set of formal topological relationships suitable for end-user interaction. In: Abel D, Chin Ooi B (eds) Advances in spatial databases, lecture notes in computer science, vol 692. Springer, Berlin, pp 277–295.http://dx.doi.org/10.1007/3-540-56869-7_16

Costamagna E (2012) Gis 3D: studio e applicazione alla documentazione dei beni culturali. Ph.D. thesis, Politecnico di Torino

Costamagna E, Spanó A (2012) Semantic models for architectural heritage documentation. Lecture Notes in computer science 7616:10

Egenhofer MJ, Franck AU (1992) Object oriented modeling for gis. URISA J IV 2:3–19

Egenhofer MJ, Herring JR (1990) A mathematical framework for the definition of topological relationships. Proceedings of the IV International Symposium on Spatial Data Handling, SDH, In

ExPath (2010) Geo Module. A geospatial API. http://expath.org/spec/geo. EXPath Candidate 30th September 2010

IAI (International Alliance for Interoperability) (2003) IFC2x2 Final Online Documentation.

ICCD (Istituto Centrale per il Catalogo e la Documentazione) (1992) Strutturazione dei dati delle schede di precatalogo : beni architettonici e ambientali, edifici e manufatti, scheda A.

ISO (International Organization for Standardization) (2006) ISO 21127:2006 Information and documentation-A reference ontology for the interchange of cultural heritage information. http://www.iso.org/iso/catalogue_detail.htm?csnumber=34424

OGC (2012) Open GIS city geography markup language (CityGML) Encoding Standard v. 2.0. Open Geospatial Consortium.

OGC (Open Geospatial Consortium) (2007) Open GIS geography markup language (GML) Encoding Standard v. 3.2.1 07–036. http://portal.opengeospatial.org/files/?artifact_id=20509

OGC (Open Geospatial Consortium) (2008) OGC KML.

OMG (Object Management Group)(2011) Unified modeling language version 2.4 - ISO/IEC 19505-2.

van Oosterom P, Jansen E, Stoter J (2005) Bridging the worlds of cad and GIS. In: Zlatanova S, Prosperi D (eds) Large-scale 3D data integration: challenges and opportunities, chap. 9, pp 9–36. Taylor and Francis-CRC Press, Boca Raton.

Peckham J, Maryanski F (1988) Semantic data models. ACM comput Surv 20, 153–189. http://doi.acm.org/10.1145/62061.62062

Repubblica Italiana (2011) "Valutazione e riduzione del rischio sismico del patrimonio culturale con riferimento alle Norme Tecniche per le Costruzioni di cui al Decreto del Ministero delle infrastrutture e dei trasporti del 14 gennaio 2008" - Direttiva del Presidente del Consiglio dei Ministri, Gazzetta Ufficiale n.47 del 26 febbraio 2011 - Supplemento ordinario n.54.

Thornes R, Bold J (1998) Documenting the cultural heritage. Getty Conservation Institute. http://archives.icom.museum/object-id/heritage/index.html

UE (Council of Europe), Granada (E) (1985) Convention for the protection of the architectural heritage of Europe. http://conventions.coe.int/Treaty/en/Treaties/Html/121.htm

W3C (World Wide Web Consortium) (2004) XML schema part 1: structures second edition, (2nd edn).

W3C (World Wide Web Consortium)(2004) XML schema part 2: datatypes second edition (2nd edn).

W3C (World Wide Web Consortium)(2008) Extensible markup language (XML) 1.0 (5th Edn).

W3C (World Wide Web Consortium)(2010) XML path language (XPath) 2.0 (2nd edn).

W3C (World Wide Web Consortium) (2010) XQuery 1.0: an XML query language (2nd edn).

Web3D Consortium (2008): Extensible 3D (X3D) ISO/IEC 19775–1:2008.

Zlatanova S, Isikdag U (2009) Towards defining a framework for automatic generation of buildings in citygml using bim. In: Zlatanova S, Jiyeong L (eds) 3D geoinformation sciences, lecture notes in geoinformation and cartography, chap 6. Springer, Berlin, pp 79–96

Towards Efficient Putrajaya 3D SDI Using Schema-Aware Compressor for CityGML

Siew Chengxi Bernad and Alias Abdul Rahman

Abstract Describing 3D city model over distributed service architecture using Open Geospatial Consortium (OGC) standard, CityGML, is getting popular e.g., Berlin-3D Spatial Data Infrastructure (SDI), Heidelberg-3D SDI, etc. Currently, visualizing 3D city model over the web using X3D is getting common where Web 3D Service (W3DS) publish 3D models in compliance with W3C X3D standard. However, involvement of 3D analysis in 3D SDI environment for Web Processing Service (WPS) requires CityGML whilst this human readable schema often raises issues such as large data that consumes high bandwidth and storage etc. Transmitting CityGML over web services without compression is unpractical due to the typical problem for CityGML. Utilizing conventional compression techniques like WinZIP, GZip, and etc., could reduce large bandwidth consumption for transmitting CityGML over the internet. However, specific schema-aware compression technique shown in this chapter could achieve even higher compression ratio and require lesser time due to utilizing the advantage of parallel programming. Moreover, compressed data is filtered and can be used for smart devices. This distinguishes the advantage using this compressor instead of conventional compression as the filtered data is relatively smaller in file size whilst maintaining essential geometries and attributes information without requiring to decompress back to original file. This chapter discusses how the technique is developed and integrated into simulated Putrajaya 3D SDI, and the results are also illustrated. We also highlight the future directions of the research in the final section.

Keywords CityGML · XML · 3D SDI · Encoder

S. Chengxi Bernad (✉) · A. Abdul Rahman
3D GIS Research Lab, Faculty of Geoinformation and Real Estate,
Universiti Teknologi Malaysia, Johor Bahru, Malaysia
e-mail: sieewster@gmail.com

A. Abdul Rahman
e-mail: alias@utm.my

A. Abdul Rahman et al. (eds.), *Developments in Multidimensional Spatial Data Models*, Lecture Notes in Geoinformation and Cartography,
DOI: 10.1007/978-3-642-36379-5_15,

1 Introduction

In 3D city domain, CityGML (Kolbe 2009), an open standard by Open Geospatial Consortium (OGC), has been implemented as GML3 application schema (Groger et al. 2008). It is commonly known that CityGML is developed as an open data model described by Extensible Markup Language (XML) schema (El-Mekawy et al. 2010). Leveraging XML standard advantages in open data sharing or data exchange serves as a solution for rapid growing web environment in terms of web services for thick or thin client–server interaction. The interoperability within 3D SDI web services could be achieved using XML-based CityGML (Basanow et al. 2008; Zipf et al. 2007). Using XML standard for data communication over web environment is used in computer science domain, such as, the instant messaging developed by Microsoft in its Window Live Messenger, etc. The nature of XML standard is a tag-based document, which allows human readable tag embedded with contents and attributes. However, such "readable" advantage could become a disadvantage in describing 3D polygonal mesh for its geometry and connectivity, since the graphic or visualization pipeline for any end user would only require geometry information such as coordinates for points, lines and polygons. The redundancies exists in CityGML due to 3D polygonal or meshes storage scheme, raises problems such as large file size that leads to storage and bandwidth consumption in smart devices within 3D SDI. This is more obvious when a large city is described in CityGML, where it could achieve several Gigabytes (GB) (Mao et al. 2010). However, due to semantic modeling and interoperability concepts, the human "readable" tag for CityGML is generally preferred.

The solution for efficient 3D data for distributed use is by employing 3D compression. The 3D compression technique used in networked application should include features such as wide applicability, high compression ratio, control over lossiness, ubiquitous access, minimum latency in reconstruction, multi-resolution representation as well as selective component-wise compression (Dinesh et al. 2002). Wide applicability defines the compression must be done for most popular representation schemes, in this case, CityGML. Furthermore, high compression ratio must be achieved as well as the control over lossiness. Ubiquitous access and minimum latency should also be followed for fast decompression with minimum overhead and latency for visualization, which is very important when models are to be visualized or used in mobile computing devices.

Replacing binary encoded data with text-based schema for communication had been discussed by an organization such as Microsoft (see MSDN 2012). Interoperability concepts encourage text-based data due to readable data storage schema. However, readability for transmitted data often comes with a cost, where large overhead of text-based data could end up producing repeated describing tags and etc. In order to maintain data interoperability while minimizing data file size during transmission, a concept of encoding within 3D SDI framework is discussed. Utilizing compression concepts such as pattern encoding, relative coding, dictionary concept as well as data referencing concepts able to reduce data size smaller than of data direct compressed by conventional method, and the results are illustrated in this chapter.

In the following chapter, some related works are presented. Furthermore, the encoding concepts are demonstrated in a workflow in Sect. 3. By using CityGML data for Putrajaya 3D Buildings, the encoding algorithm is tested and the results are tabled and discussed in Sect. 4. In the final section of this chapter, some issues are highlighted and discussed for further experiments and directions.

2 Related Works

Various SDI has been proposed and established e.g., Infrastructure for Spatial Information in the European Community (INSPIRE), Digital China Geospatial Framework (DCGF) (Lia et al. 2008), National Land Information System (NLIS) in Poland (Gazdzicki and Linsenbarth 2004) etc., with the aim to make geospatial data exchangeable for supplying geospatial information. On the other hand, 3D SDI for city domain is getting popular as seen in Heidelberg 3D-SDI (Basanow et al. 2008), Berlin-3D SDI, and etc. This chapter only discusses 3D SDI and how the existing 3D SDIs publish spatial information over the web environment.

In existing SDIs, the infrastructures are generally built on OGC standard web services, such as Web Feature Services (WFS), Web Map Services (WMS), Web Coverage Services (WCS) and etc. These web services functionalities and specification details are found in OGC specification documents. On top of that, Heidelberg 3D SDI and Berlin 3D SDI incorporate Web 3D Service (W3DS) on top of WFS layer and uses X3D (see X3Dom 2012) as 3D portrayal pipeline standard. 3D Symbology is described by 3D Style Layer Descriptor (SLD3D), which is an extension of 2D Style Layer Descriptor. In Heidelberg 3D SDI, Open Location Service (OpenLS) (Schilling et al. 2009) is introduced to enable routing service over 3D SDI environment (Fig. 1).

On the other hand, the snapshot in Fig. 2 depicted a framework from 3D Portrayal Interoperability Experiment (3DPIE) which is an initiative to establish a 3D portrayal standard for 3D SDI. Usage categories are defined as thin client—thick server, medium client—medium server, and thick client—thin server. Retrieval of 3D data will be presented in X3D format using W3DS. The portrayal standard depicted the interaction between web services for 3D data portrayal.

One interoperable example discussed in the dissertation of Galip (2007) that various data models are bridged and interoperable achieved by employing Geographic Markup Language (GML) as the intermediate data exchange standard. Furthermore, a few XML infoset techniques are used to encode large dataset generated by streaming data such as earthquake and some other sensors data. The encoding techniques used are tested in the distributed environment where encoding engine is used within web services as depicted in Fig. 3.

In Fig. 3, it is obvious that general binary XML is used without modification, where encoding process could be integrated and deployed. Geospatial data is

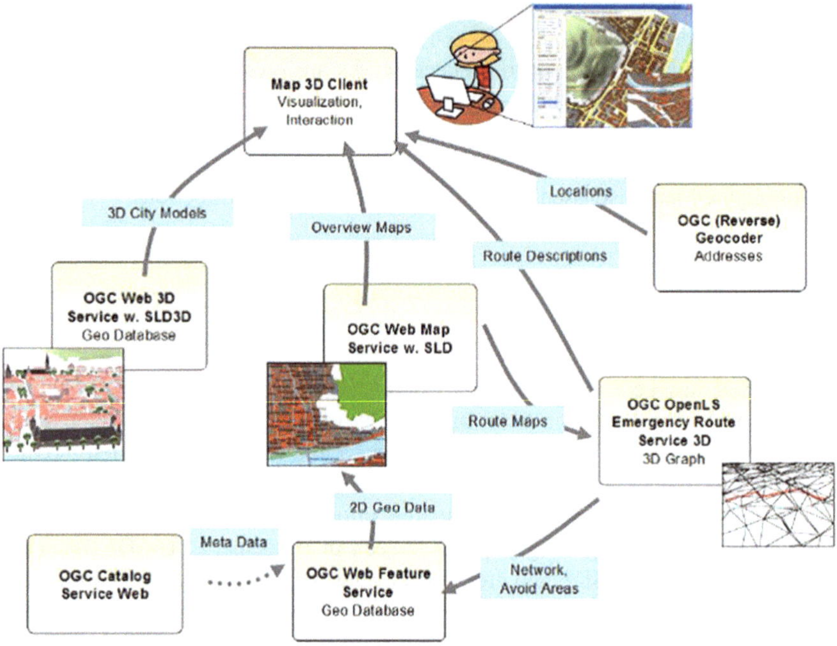

Fig. 1 Heidelberg 3D SDI (Basanow et al. 2008)

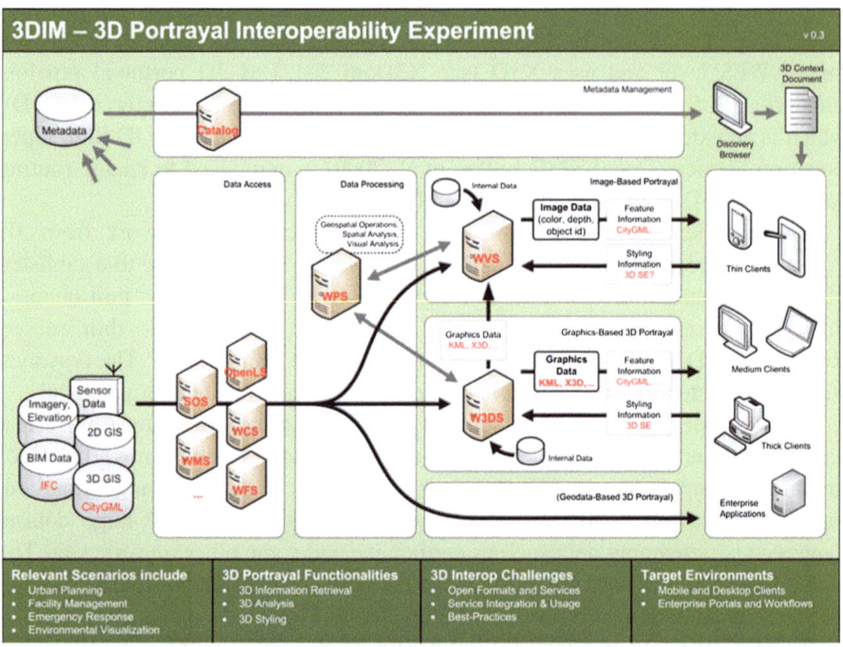

Fig. 2 3D Portrayal interoperability experiment framework (Schilling and Kolbe 2010)

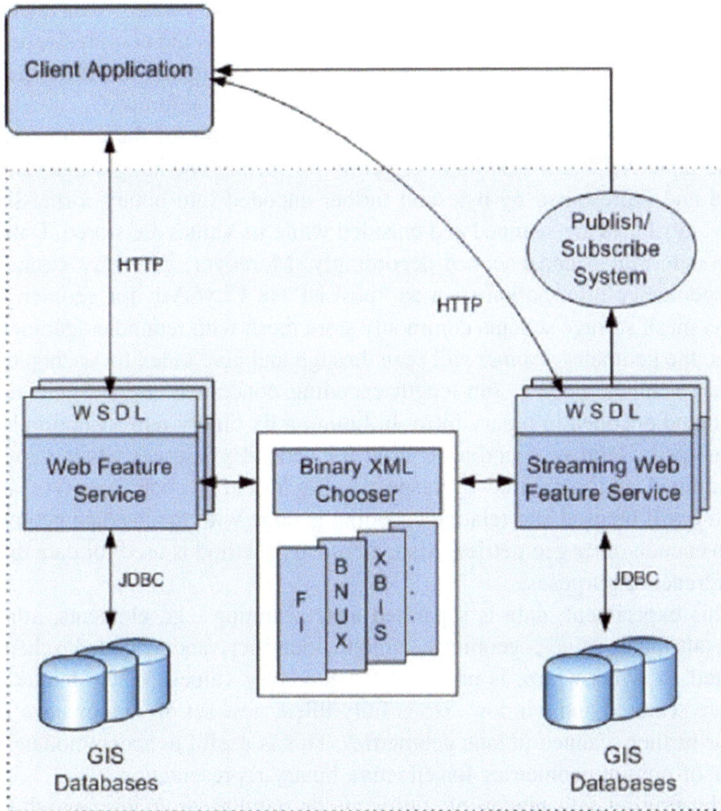

Fig. 3 Encoding engine used in distributed services

transferred within web services encoded by binary XML and decoded for further usage e.g., visualization or analysis. This example shows the possibility of integration for encoding engine within web services. However, the data used in this example is purely GML while in 3D data, vertices and attributes are much redundant compared to pure GML data.

3 Schema-Aware Compressor for CityGML

Conventional compression methods for text data are useful in common scenarios. Commonly the techniques are developed to solve general data encoding, where performance is limited as these methods could not serve best for all types of data. A CityGML schema-aware compressor (CitySAC) is developed to specifically encode CityGML by employing general compression concepts.

This compressor is developed with preset "schema-aware" data dictionary, where tags are well "understood" by the compressor, as the compressor used the CityGML schema as its dictionary for tags matching purposes. Furthermore, XSD information is also used to encode data based on "type-reference".

When CityGML is used as input, the scanner module of the compressor will scan the input XSD and used its information for further encoding usage. Tags will be used and represented by byte and further encoded into binary form. Simultaneously, attributes are scanned and encoded while its values are stored. Data types will be determined and encoded accordingly. Moreover, geometry scanner will search geometry information such as "posList" in CityGML for geometry mapping. As mesh storage schema commonly store mesh with redundant geometries at vertices, the geometry scanner will scan through and give index for each geometry.

In tag scanner module, run-length encoding concept is used, where tags are scanned and encoded in binary form. In future, a fix binary representation list will be developed. Relative encoding is used for geometry scanner where geometries are examined and compared by respective X, Y and Z coordinate. An average geometry will be used and relative encoding is done with the average geometry as basis to encode other geometries. Also, dictionary method is used for data mapping and referencing purposes.

In this experiment, data is separated after scanning e.g., elements, attributes, values, attribute values, geometry, unique identifier, and encoded scheme are generated. A window size is provided for geometry values. When unique geometries are scanned and window size is fully filled, new set of window size will be used for further scanned unique geometries. This is useful to accommodate higher number of unique geometries for effective binary representation.

By leveraging advantages of multicore or parallel programming, this compressor is developed with parallel processing capability where tasks are divided into multi-cores depending on running machine. The efficiency of multicore is largely seen in next chapter. Figure 4 depicted the workflow of the compressor.

4 Results and Discussions

In this experiment, several datasets were used as input e.g., several different Putrajaya 3D buildings. This data is simulated by Google Sketch-Up in CityGML 1.0 standard compliant. These datasets are tested with the in-house developed compressor CitySAC, Winzip Deflate, Fast-Infoset, LZMA and Bzip2. Due to possible quadratic time complexity issue raised by Coors and Ewald (2005), topology scanner is excluded from the workflow and replaced by pattern encoding. Rescanning topology encoded in CityGML could end up time-consuming for data up to and beyond Level of Detail-3 (LoD-3). Unless connectivity compression is necessarily required in the workflow, the topology scanning for geometry information incurred heavy time consumption and defeats the overall objective. The results are shown in Table 1. Datasets of Putrajaya in LoD-3 is used in the

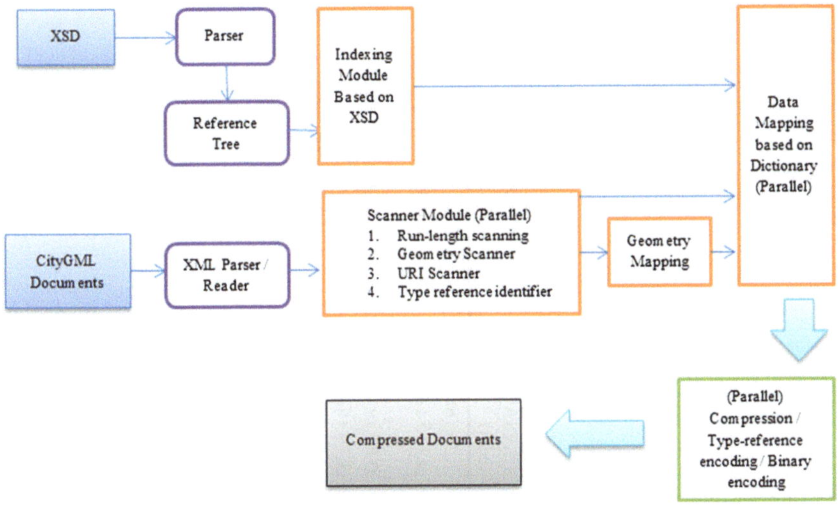

Fig. 4 The CitySAC workflow in modules

experiment and is shown in various range of original file size e.g., from less than 1 MegaByte (MB) to datasets more than 20 MB.

Meanwhile, the timespan is shown in Table 2.

The timespan in line graph shown in Fig. 5.

With Deflate Zip being the benchmark basis, LZMA, BZip2 and CitySAC—LZMA is compared with the respective results of average improvement of 17.49 % (BZIP2), 31.96 % (LZMA), and 37.78 % (CitySAC + LZMA). This shows that CitySAC method improved the most compared to others techniques. Also in Fig. 5, CitySAC timespan achieved average result while LZMA is going exponential when file size is getting bigger.

The experiment is done by creating a standalone package developed using Microsoft C# Visual Studio platform. The nature of C# requires DOT NET Framework in order to debug or publish the package. Low level language such as C++ or C could improve the timespan of compression due to the nativity in computer environment. Using C# library to construct variables in run-time would consume more time compared to pointers which could be created in those low level languages and could largely reduce the compression time consumption. Though binary encoding framework such as Fast-Infoset is useful in most cases, however, in our experiment, the compression ratio is not satisfying where only 50–60 % is achieved, while CitySAC achieved at least 91 %. The integration of the compressor into distributed service architecture is illustrated in Fig. 6.

In this experiment, decompressed data is a type of filtered data, while the de-compressor also provides option to generate back to original file. Flexibility is gained in this procedure where CitySAC provides options, where one of which is query-able while common decompression technique discussed in Table 1 will only generate the original file. In Fig. 6, XML document database that store 3D data as

Table 1 The comparison in terms of file size for different techniques

	Original (MB)	Deflate zip (MB)	Fast infoset (MB)	LZMA (MB)	Bzip2 (MB)	CitySAC + Deflate (MB)	CitySAC + LZMA (MB)
Putrajaya_Mosqu e.xml	10.928	1.56	4.81	0.992	1.226	1.23 (88.7 %)	0.88 (91.95 %)
Putrajaya Convention.xml	0.962	0.17	0.43	0.102	0.106	0.115 (87.9 %)	0.090 (90.65 %)
National Audit.xml	8.594	0.948	4.132	0.696	0.869	0.87 (89.9 %)	0.659 (92.34 %)
SeriGemilangBrid e.xml	27.454	3.612	12.352	2.714	3.519	3.29 (88.1 %)	2.53 (90.79 %)

Table 2 The timespan taken for each technique

Input\techniques	Deflate z ip (s)	Fast- Infoset (s)	LZMA (s)	Bzip2 (s)	CitySAC + deflate (s)	CitySAC + LZMA (s)
Putrajaya Mosque.x ml	3.8	4.2	5.8	2.2	3.1	4.1
Putrajaya Convention.xml	1.1	0.9	1.3	0.3	0.42	0.51
National Audit.xml	3.8	3.6	4.5	1.2	1.6	1.9
SeriGemilangBride.x	9.8	11.5	15.2	4.5	10.5	11.2

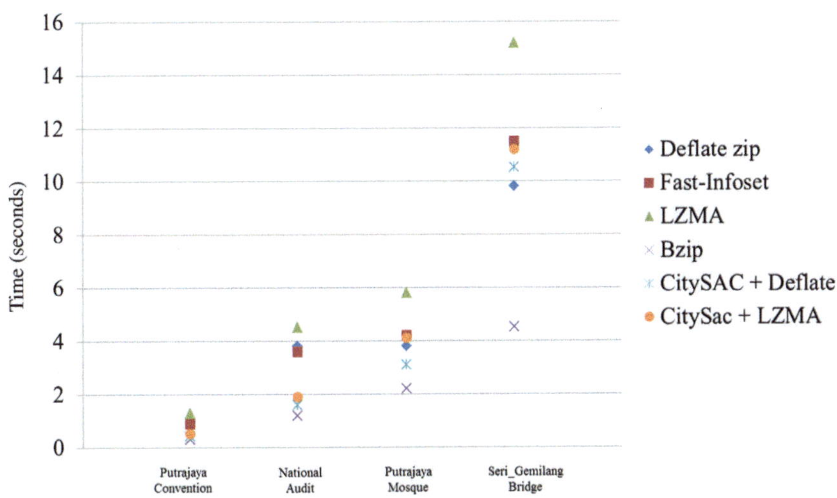

Fig. 5 The timespan at vertical axis depicted comparing different techniques

geodatabase will be published based on query to web services. Integration of encoder and decoder in the framework allow encoding and decoding of data over distributed services. The compressor could be used especially when full CityGML content is required e.g., 3D analysis for particular web service. The compressor should facilitate the streaming process in binary encoded form. Other than visualization, compressed CityGML could be used for various scenarios including medium client analysis. By allowing the option to encode data in SDI environment, web services will be developed to test the efficiency in streaming environment.

5 Concluding Remarks

The compressor is developed for specific data such as CityGML. This chapter discusses the compression technique for CityGML LoD-3 by employing specific schema-aware technique as well as scanning module utilizing general compression

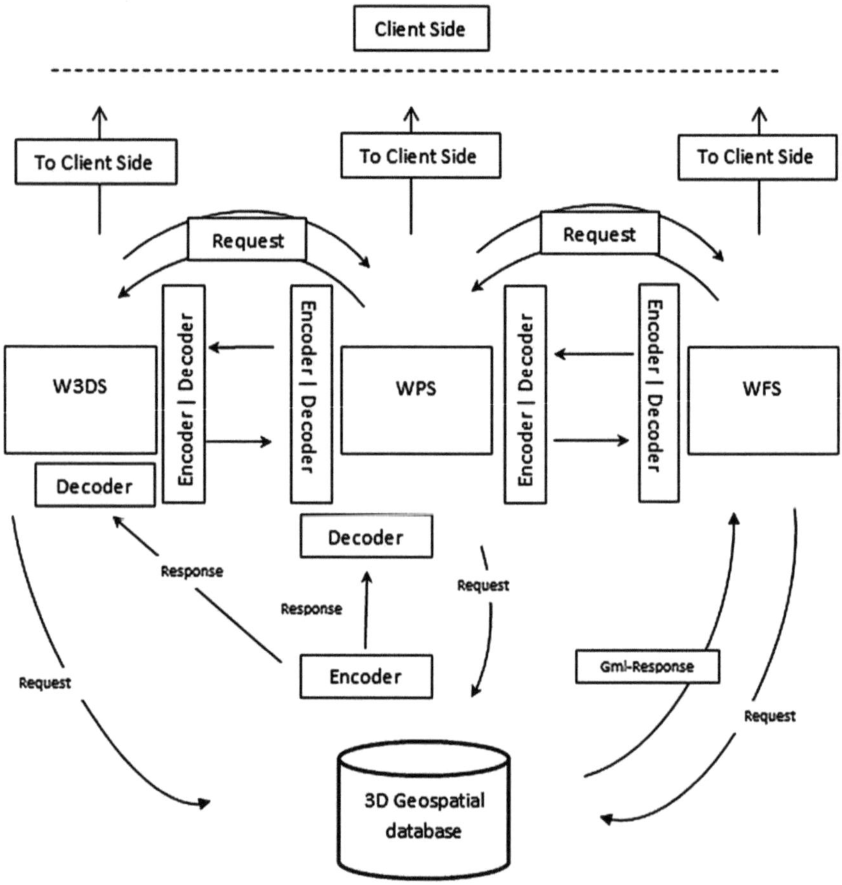

Fig. 6 The integration of encoding engine into 3D SDI environment

concepts. Since CityGML is now the standard for 3D geospatial data sharing, the encoding of CityGML could be seen useful in complex web service environment. In this research, better efficiency is achieved by taking total time consumption for transmission and comparing compression ratio with different techniques. Specific schema-aware compressor shows the scenarios where XML compression before feeding the data into conventional compressor, could achieve faster and higher compression ratio e.g., 90 % based on the datasets in LoD-3. A specific retrieving procedure could be developed to read compressed data without requiring decompressing the entire content. The transactions of atomized data will be defined and tested in web service environment. Currently, the de-compressor publishes CityGML from compressed data. In future, a mobile viewer will be developed to directly visualize compressed data.

Acknowledgments We would like to convey our deepest acknowledgement to Ministry of Higher Education (MOHE), Malaysia for the scholarship under the program MyPhD, MyBrain15 and enabling us to carry out this research project.

References

Basanow J, Pascal N, Neubauer S, Schilling A, Zipf A (2008) Towards 3D spatial data infrastructures (3D-SDI) based on open standards—experiences, results and future issues. In: Oosterom P et al (eds) Advances in 3D geoinformation systems. Springer, Heidelberg, pp 65–86

Coors V and Ewald K (2005) Compressed 3D urban models for internet-based e-planning. In: Proceeding of 1st international workshop on next generation 3D city models. EUROSDR Publication #49, Bonn

Dinesh S, Babji SV, Mudur SP (2002) Compression techniques for distributed use of 3D data: an emerging media type on the internet. In: ICCC '02 Proceedings of the 15th international conference on computer communication. Washington, pp 676–696

El-Mekawy M, Ostnam A, Shahzad K (2010) Towards interoperating CityGML and IFC building models: a unified model based approach, advances in 3D geo-information sciences LNG&C. Springer, Berlin, pp 73–94

Galip A (2007), Service oriented architecture for geographic information systems supporting real time data grids. Dissertation, Department of Computer Science Indiana University, p 238

Gröger G, Kolbe TH, Czerwinski A, Nagel C (2008) OpenGIS® city geography markup language (CityGML) encoding standard, version: 1.0.0, OGC 08-007r1, p. 234

Gazdzicki J, Linsenbarth A (2004) 10th EC GI & GIS workshop. ESDI State of the Art, Warsaw, Poland

Kolbe TH (2009) Representing and exchanging 3D city models with CityGML. In: Lee J, Zlatanova S (eds) 3D geo-information sciences: LNG&C. Springer, Berlin/Heidelberg, pp 15–31

Lia P, Lan W, Xiao X (2008) SDI in China: progress and issues. In: The international archives of the photogrammetry, remote sensing and spatial information sciences, vol XXXVII, Part B4. Beijing

Mao B, Ban Y, Harrie L (2010) A multiple representation data structure for automatic visualization of generalized 3D city models, geoinformatics, KTH, Drottning Kristinas, vag 30. Lund University, Sweden

MSDN (2012) http://msdn.microsoft.com/en-us/library/ms733742.aspx. Accessed 6 Aug 2012

Schilling A, Kolbe TH (2010) Draft for candidate OpenGIS® web 3D service interface standard, version: 0.4.0, OGC 09-104r1, p. 95

Schilling A, Over M, Neubauer S, Neis P, Walenciak, Zipf A (2009) Interoperable location based services for 3D cities on the web using user generated content from OpenStreetMap. UDMS

X3Dom (2012) http://x3dom.org. Accessed 5 May 2012

Zipf A, Basanow J, Neis P, Neubauer S, Schilling A (2007) Towards 3D spatial data infrastructures (3D-SDI) based on Open Standards—experiences, results and future issues. In: "3D GeoInfo07". ISPRS WG IV/8 international workshop on 3D geo-information: requirements, acquisition, modelling, analysis, visualisation. Delft, Netherlands

MIX
Papier aus verantwortungsvollen Quellen
Paper from responsible sources
FSC® C105338

If you have any concerns about our products,
you can contact us on
ProductSafety@springernature.com

In case Publisher is established outside the EU,
the EU authorized representative is:
Springer Nature Customer Service Center GmbH
Europaplatz 3, 69115 Heidelberg, Germany

Printed by Libri Plureos GmbH
in Hamburg, Germany